Recent Results in Cancer Research

Fortschritte der Krebsforschung

Progrès dans les recherches sur le cancer

20

Edited by

V. G. Allfrey, New York · M. Allgöwer, Basel · K. H. Bauer, Heidelberg · I. Berenblum, Rehovoth · F. Bergel, Jersey · J. Bernard, Paris · W. Bernhard, Villejuif N. N. Blokhin, Moskva · H. E. Bock, Tübingen · P. Bucalossi, Milano · A. V. Chaklin, Moskva · M. Chorazy, Gliwice · G. J. Cunningham, Richmond · W. Dameshek, Boston M. Dargent, Lyon · G. Della Porta, Milano · P. Denoix, Villejuif · R. Dulbecco, La Jolla · H. Eagle, New York · R. Eker, Oslo · P. Grabar, Paris · H. Hamperl, Bonn R. J. C. Harris, London · E. Hecker, Heidelberg · R. Herbeuval, Nancy · J. Higginson, Lyon · W. C. Hueper, Fort Myers · H. Isliker, Lausanne · D. A. Karnofsky, New York · J. Kieler, København · G. Klein, Stockholm · H. Koprowski, Philadelphia · L. G. Koss, New York · G. Martz, Zürich · G. Mathé, Villejuif · O. Mühlbock, Amsterdam · W. Nakahara, Tokyo · V. R. Potter, Madison · A. B. Sabin, Cincinnati · L. Sachs, Rehovoth · E. A. Saxén, Helsinki · W. Szybalski, Madison H. Tagnon, Bruxelles · R. M. Taylor, Toronto · A. Tissières, Genève · E. Uehlinger, Zürich · R. W. Wissler, Chicago · T. Yoshida, Tokyo

Editor in chief

P. Rentchnick, Genève

Springer-Verlag Berlin Heidelberg GmbH 1969

Rubidomycin

A New Agent against Cancer

Edited by

J. Bernard · R. Paul · M. Boiron
Cl. Jacquillat · R. Maral

With 68 Figures

Springer-Verlag Berlin Heidelberg GmbH 1969

Sponsored by the Swiss League against Cancer

ISBN 978-3-642-88127-5   ISBN 978-3-642-88125-1 (eBook)
DOI 10.1007/ 978-3-642-88125-1

Title No. 3635

List of Authors

Institut de Recherches sur les Leucémies et les Maladies du Sang de l'Université de Paris — 2, Place du Docteur Fournier — 75 — Paris Xe — France

Directeur: Professeur JEAN BERNARD

A. BASCH, J. L. BINET, M. BOIRON, J. BONHOMME, A. BUSSEL, I. CAEN, J. CHASSIGNEUX, F. CHAVELET, J. DAUSSET, J. DELOBEL, C. DRESCH, J. DUMONT, J. M. DUPUY, F. EBERLIN, G. FLANDRIN, A. GOGUEL, CL. JACQUILLAT, F. M. KOURILSKY, M. J. LARRIEU, A. LEVACHER, D. LEVY, J. P. LEVY, P. LORTHOLARY, CL. MACREZ (Service de Cardiologie — Hôpital Saint-Louis), H. MARNEFFE-LEBRESQUIER, Y. NAJEAN, F. MIELOT, J. RIPAULT, G. SCHAISON, M. SELIGMANN, Y. SULTAN, J. TANZER, M. THOMAS, F. TEILLET, M. WEIL, C. WEISGERBER

Société des Usines Chimiques Rhône-Poulenc

Directeur Scientifique: RAYMOND PAUL
Laboratoires de Recherches — 94 — Vitry-Sur-Seine
Directeur: P. VIAUD

R. DESPOIS, R. JACOB. — O. LEAU. — A. BELLOC. G. BOURAT, Y. CHARPENTIE, N. DE CHEZELLES, M. DUBOST, R. DUCROT, J. FOURNEL, P. GANTER, D. JUGE, L. JULOU, F. KOENIG, J. LUNEL, D. MANCY, R. MARAL, F. MOTTET, J. MYON, L. NINET, S. PASCAL, J. PASQUET, S. PINNERT, P. POPULAIRE, J. PREUD'HOMME, Y. DE RATULD, J. RENAUT, G. H. WERNER

Istituto Ricerche Farmitalia, Via dei Gracchi, 35, Milano/Italia

Directeur: B. CAMERINO
F. ARCAMONE, A. DI MARCO, C. SPALLA

P. BARBIERI, O. BELLINI, G. BORETTI, E. CALENDI, G. CANEVAZZI, G. CASSINELLI, T. DASDIA, L. DORIGOTTI, A. FIORETTI, G. FRANCESCHI, M. GAETANI, A. GREIN, R. MAZZOLENI, M. MENOZZI, P. OREZZI, A. RUSCONI, A. SANFILIPPO, B. SCARPINATO, T. SCOTTI, R. SILVESTRINI, M. SOLDATI, L. VALENTINI

Institut für experimentelle Pathologie der Farbenfabriken Bayer AG., Wuppertal-Elberfeld, Bundesrepublik Deutschland

Directeur: E. GRUNDMANN
R. BIERLING, H. J. SEIDEL

Contents

Part I — Preparation and Experimental Investigation

Part II — Clinical and Therapeutic Study

Introduction

Trials in the treatment of the leukemias are sometimes based on a hypothesis, as in the case of exchange transfusion [33] or the use of antimetabolites [86]. Or they are conducted empirically as the results of chance observations, as in the case of the use of the nitrogen mustards, urethane, and the *Vinca* alkaloids. Or they lie between the two, aiming at making use of well established biological facts: examples are the use of ACTH, cortisone and, more recently, the antibiotics.

What is true for bacteria may also perhaps be true, if not for elephants, at least for the malignant cells of mammals. It was this idea that lay behind the first attempts at treating cancer and leukemia with antibiotics. The results obtained by the use of certain substances extracted from micro-organisms (actinomycin, azaserine, mitomycin, rufocromomycin), although encouraging at times, were inconsistent. The action of rubidomycin appears to cover a wider spectrum and to be more consistent and more effective.

As in the case of Homer and Christopher Columbus, the honor of having given birth to rubidomycin is claimed by more than one country and town. In fact, the same product was discovered in the same year, though quite independently, by a group of French workers [184] who described it under the name of rubidomycin, and by a group of Italian workers [87] who studied it under the name of daunomycin. Shortly afterwards, an exchange of communications between these two groups showed that the product was one and the same.

The first clinical trials with rubidomycin were carried out in Paris [15, 131] and with daunomycin in New York [213].

As a result of a recent symposium held at the Hôpital Saint-Louis at Paris, it is now possible to assess this drug [58].

Rubidomycin can produce complete remission in acute leukemia, both in new cases and in cases where it has become resistant to other drugs. It differs from other drugs for the treatment of acute leukemia in three respects, the first advantageous, the second doubtful, and the third disadvantageous.

It has a very wide spectrum of activity, being effective against both lymphoblastic leukemia and myeloblastic leukemia. In this, it is unusual. In our first trials with it we obtained complete remission in 60% of cases of lymphoblastic leukemia refractory to the treatment previously used, and complete remission in 50% of cases of myeloblastic leukemia. Rubidomycin also appears capable of sometimes influencing for the better the course of a wide range of diseases such as reticulosarcoma and chronic myeloid leukemia.

Its action is both very rapid and very powerful. The number of white cells per cubic millimeter may fall in a few days from 200 000 to 200. This is a major disadvantage, since profound aplasia of the bone marrow often occurs at early stage; but it has to be accepted and vigorous symptomatic treatment has to be applied, which is often effective. Transfusions of platelets or of leukocytes of chronic myeloid leukemia at the myelocytic stage or appropriate antibiotic therapy may be needed. The rapid action is also a great advantage when it is necessary quickly to arrest the course of a rapidly developing leukemia. A way must be picked, often with difficulty, between doses that are too low and have, in certain cases, been the cause of failure and doses that are too high and result in irreversible aplasia. The great capacity of rubidomycin for causing aplasia, and perhaps the unusual way in which it achieves its effects, have renewed interest in the regular cytological study of leukemias under treatment. Treatment is sometimes facilitated, sometimes complicated thereby. It is facilitated when the changes in the shape and size of the leukemic cells and the appearance of gigantoblasts provide the first signs of the drug's activity, thus making it easier to control the treatment than was the case with the earlier antileukemic drugs. It is complicated when it is necessary to decide whether medullary leukoblastosis subsequent to aplasia indicates a rapid relapse or instead incomplete recovery with disturbances of maturation due to the rubidomycin.

Rubidomycin is directly or indirectly toxic to the heart and, in certain cases, has caused fatal myocardial insufficiency, but apparently, only or particularly, after prolonged courses of treatment with a high total dose of the drug. As things are at present, it would be wise not to give maintenance treatment over long periods, not to give the drug to cardiac patients, and to keep all patients under constant clinical surveillance and electrical and radiological monitoring.

What does the future hold out for those patients who have been treated with rubidomycin or, to be more accurate, what does the future hold out for rubidomycin itself? Will it be failure or oblivion after a short period of success, the beginning of a major advance or, more modestly, its establishment as a useful drug with precise but limited indications? There have been so many disappointments during the past twenty years that any forecast must necessarily call for caution. Nevertheless, whether looking at the present or taking a short-term or long-term view, the outlook seems to be promising. Two things now appear certain: rubidomycin can cause remissions in lymphoblastic leukemia refractory to other forms of treatment, and it can cause remissions in 50% of cases of myeloblastic leukemia. Thus two of the most serious gaps in treatment may be at least partially filled.

The investigation of combinations of rubidomycin with other antileukemic agents by modern chemotherapeutic methods is only just in its infancy. It is being introduced along with vincristine, prednisone, methotrexate, and 6-mercaptopurine into the systematic treatment of lymphoblastic leukemias with the aim of extending the period of complete remission. It is being employed for the treatment of myeloblastic leukemia along with other drugs such as methyl-glyoxal-bis-guanyl hydrazone and cytosine arabinoside, with the aim of increasing the number of complete remissions. These would all seem to be promising possibilities.

More striking advances may be made possible either by the association of this potent but perhaps rather short-acting drug with some form of immunotherapy or by the development of derivatives less toxic to the heart, or by the discovery that the action of other antibiotics is greatly stimulated by the effects of rubidomycin, or by the discovery of new groups of drugs of hitherto unknown modes of action. It is very difficult at the present time to venture upon any forecast of this kind.

History of Rubidomycin

Although short, the history of rubidomycin can be divided into three periods.

The First Antimitotic Antibiotics

The first step in the research that was to lead to the discovery of rubidomycin was taken by WAKSMAN and WOODRUFF [227], who isolated actinomycin in 1940. The actinomycins, which are antibiotics extracted from the actinomycete *Streptomyces antibioticus,* have the capacity of preventing the development of various tumors, both experimental and human [42, 110, 111, 226].

Great efforts were made in the years that followed this discovery, and new antimitotic antibiotics were isolated. The results of all these efforts were both remarkable and disappointing. They were remarkable because the effects of some of the substances isolated could be studied with great precision thus placing some of the antibiotics among the most useful tools of molecular biology. They were remarkable too because certain human tumors such as Wilms' tumor and choriocarcinoma have been found to be very sensitive to their action. On the other hand, they were disappointing in that the number of sensitive human tumors is extremely limited and the action of the antimitotic antibiotics is incomplete and inconsistent. In the field of the leukemias and related diseases, the results were particularly modest. The Japanese drug mitomycin has a certain effect on chronic myeloid leukemia, but it is much less effective than the drugs generally employed, such as busulfan [112]. Rufocromomycin [130] has been investigated by our group and has been shown to produce remissions in certain cases of lymphosarcoma and reticulosarcoma refractory to other forms of treatment, but it is essentially no more than of secondary value.

Isolation of Rubidomycin

The isolation and study of rubidomycin transformed the antibiotic treatment of leukemia, which until then was in a highly unsatisfactory state.

Particular attention was given to the study of the anthracycline group from 1950 on [41].

In 1962, a group of French chemists [184] described three substances isolated from *Streptomyces caeruleorubidus,* the most interesting of which became known as rubidomycin. Its properties and its action on various experimental tumors were established [79].

Work was also carried out in this field in Italy and the Soviet Union. The same substance was isolated in Milan from *Streptomyces peucetius* and called daunomycin [68]. The substance rubomycin [98, 99] was prepared in Moscow, its C fraction being identical with rubidomycin and daunomycin. Research on rubomycin has been carried out in the Soviet Union [12, 77, 102, 105, 158, 206], and its antitumor action has been demonstrated [160, 161, 187, 198, 199].

The First Clinical Trials

The first clinical trials with rubidomycin were carried out at the Hôpital Saint-Louis in Paris in 1965. They were at the outset very cautious. For ethical reasons, the only cases treated with the new drug were of very serious acute leukemia refractory to other forms of treatment. Encouraging results were quickly obtained and the remarkable effect of rubidomycin on the course of leukemia very rapidly acknowledged. A large-scale trial was undertaken at the beginning of 1965, and more than 800 patients were treated with rubidomycin between 1965 and 1968.

Curiously enough, daunomycin, though discovered in Italy, does not seem to have been employed, at least on a large scale, by Italian physicians and was not, at any rate at the outset, used in the treatment of leukemia. The first trials were made in the United States in the treatment of solid tumors, with only limited success, but a large-scale clinical trial was soon undertaken by various teams at the Sloan Kettering Institute, particularly by J. BURCHENAL and CHARLOTTE TAN.

Although satisfactory results were obtained both by the Chemotherapy Unit of the Leukemia Research Institute of the Hôpital Saint-Louis and by the Sloan Kettering Institute, the results obtained by other groups were very inconsistent, probably because the doses administered were too small.

The First Overall Surveys

The comparison and correlation of the results obtained in these first trials were plainly necessary. They formed the chief topic of the first international symposium on rubidomycin and daunomycin, held in Paris in March 1967 [58]. This symposium was attended by European and American workers who had studied the new drug and those who had employed it, and it included chemists, pharmacologists, pathologists, and clinicians.

The essential purpose of this monograph is to present an overall survey of rubidomycin. It is based partly on the results obtained by the various teams that have used the drug throughout the world but also partly, naturally enough, on our own personal experience with it.

Part I — Preparation and Experimental Investigation

Chapter 1

Preparation

Rubidomycin [1], obtained from cultures of *Streptomyces coeruleorubidus* [63, 79, 80, 184], is a basic antibiotic, orange-red in color, belonging to the group studied by OLLIS and SUTHERLAND [175]. This group of antibiotics, to which the name of anthracyclines was given by BROCKMANN [40], includes among others aklavin [208], the cinerubins [84], nogalamycin [223], the pyrromycins [43], the rhodomycins [41], the isorhodomycins [44], and rutilantin [8], most of which possess cytostatic properties. Rubidomycin is identical with daunomycin, an antibiotic [2] obtained from cultures of *Streptomyces peucetius* [52, 68, 69, 71]. This has recently been confirmed by TONG et al. [219 bis].

1. Production

a) Productive Strains

The first productive strains, described in 1962 [184], were obtained from samples of soil from various sources. They belong to the species *Streptomyces coeruleorubidus*, originally isolated by GAUSE et al. [101]. Rubidomycin production appears to be fairly widespread in this species; GAUSE in fact showed that a representative of the species which he had described produced a mixture of metabolites forming rubomycin, the C fraction of which is identical with rubidomycin [99, 100, 102, 180].

Streptomyces coeruleorubidus is characterized in particular by a vegetative mycelium of a color ranging generally from pink to orange-pink and up to brownish red, passing through numerous intermediate shades depending on the culture medium. The aerial mycelium is turquoise blue in color when the spores are formed. In most culture media the various strains secrete a soluble pigment, the color of which ranges from pink through orange-red to brownish red, depending on the nature of the

---

[1] Rubidomycin: daunorubicin international generic name; Cérubidine Specia.

[2] "The two companies Rhone-Poulenc and Farmitalia independently isolated in their research laboratories an antimitotic antibiotic which they called 'rubidomycin' and 'daunomycin' respectively before further research showed that it was one and the same product. At the present time, both companies confirm that the products supplied to clinicians for trials under those names meet the same physicochemical and biological standards and have the same activity." Presse méd. **75**, 1381 (1967).

culture medium and its acidity or alkalinity. In addition, on a suitable medium containing tyrosine, *S. coeruleorubidus* produces a melanic pigment of an intense black color, so that the species must be included among those which produce melanin.

Rubidomycin can also be produced by other species, in particular *Streptomyces peucetius*, described by GREIN and SPALLA [108], which produces daunomycin [109]. This species differs from *S. coeruleorubidus*, on the one hand by the formation of a pinkish white aerial mycelium and inability to produce a melanic pigment, on the other by the shape of the fruiting spores, which end in hooks or curls but not spirals, whereas the fruiting spores of *S. coeruleorubidus* form tightly coiled spirals containing several turns.

b) Production in Fermenters

Rubidomycin is produced by the classic procedures of submerged fermentation. The first cultures are made in an agar medium, inoculated by means of spores. The cultures are then propagated in a shaken liquid medium, in apparatus of increasing size ranging from 2-liter Erlenmeyer flasks to fermenters several cubic meters in volume. In addition to starch and soya oil, the culture medium contains various nutritive substances such as soya flour and distillers' solubles. The maximum concentration of rubidomycin is obtained after four days culture, and reaches several tens of milligrams per liter. For this reason the broth is pink in color.

2. Isolation

The active substances are first extracted from the broth in the form of a crude base containing a mixture of rubidomycin and other substances. The constituents of this crude base are then fractionated and purified by counter-current distribution. The rubidomycin is then crystallized out as the hydrochloride.

a) Extraction of the Crude Base

The broth is acidified to pH 1.5 by means of a concentrated solution of oxalic acid; a filter aid is then added and the broth filtered. The filtrate, partially neutralized to pH 4.5, is passed through column containing a carboxylic ion-exchange resin (Amberlite IRC-50) in the form of the free acid. After all the filtrate has passed through the column, it is washed with water and then with methanol containing 10% of water. Elution is effected by means of a solution of methanol with 10% of water containing 10 g/liter of sodium chloride. The middle fraction, which is orange-red in color, is collected and then evaporated down to small volume under reduced pressure.

The pH of the concentrate is adjusted to a value of 7—8, and the crude base is extracted with chloroform. Concentration of the chloroform solution under reduced pressure, followed by precipitation by means of hexane, gives a mixture of rubidomycin with other substances.

The crude base thus obtained is partially purified by solution in a mixture of methylene chloride and carbon tetrachloride in the proportion of 5:1 by volume and by extraction of the active products in water by acidification to pH 3. The aqueous phase is then repeatedly extracted with methylene chloride, the pH being adjusted to 7.8. The solution in methylene chloride is concentrated and precipitated by means of hexane, which gives a semi-purified base formed of a mixture of the A, B, and C constituents.

With this procedure, the processing of 520 liters of broth gives a yield of about 5 g of semi-purified base, brick red in color, and of approximately the following composition:

Constituent A (13057 R.P.) rubidomycin approx. 35% by weight
Constituent B (13213 R.P.) approx. 20% by weight
Constituent C (13330 R.P.) approx. 10% by weight

b) Fractionation by Counter-current Distribution

The semi-purified base then undergoes 100 transfers by counter-current distribution in an automatic apparatus (Quickfit type 210/25 F) consisting of 100 cells each of 50 ml capacity. The solvent system used consisted of two separate phases of the mixture of ethyl acetate, methanol, and M/15 phosphate buffer at pH 5.6 in the proportions 8:3:5 by volume. Ten g of base were dissolved in 200 ml of one 1:1 mixture of the two phases and the solutions were divided up between the first four cells of the apparatus; counter-current distribution was then carried out by the classical procedure until 100 transfers had been effected. The distribution curve was then drawn from the absorption at 490 nm of the contents of each of the cells (after homogenization by dilution of aliquot parts of the two phases with methanol). The curve (Fig. 1) shows the presence of three peaks corresponding to the partition

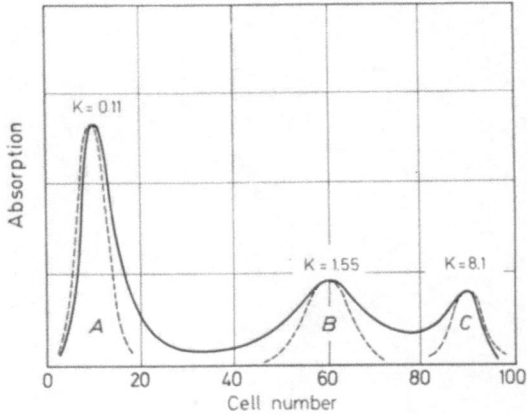

Fig. 1. Counter-current distribution of the crude base. ———— Experimental curve;
- - - - - Theoretical curve

Table 1. *Fractionation by counter-current distribution*

Con-stituent	Cell number corresponding to peaks	K = partition coefficient	Cell number containing a pure constituent	Weight of base isolated in g
A	10	0.1	4—15	2.9
B	61	1.55	52—68	1.2
C	89	8.1	85—94	0.7

coefficients indicated in Table 1. The solute content of the cells containing the separate constituents was determined from a comparison between the experimental curve and the theoretical distribution curves calculated from the position of the peaks. The solutions from the cells containing the pure constituents were collected, and the corresponding bases isolated by solvent extraction at pH 7.5, concentration, and precipitation by hexane. The results of the fractionation are shown in Table 1.

Chromatographic analysis of these bases showed that constituent A was pure. Constituent B contained small amounts of constituent A and an inactive red substance (aglycone). These impurities are due to the decomposition of constituent B during isolation, a phenomenon which was demonstrated by repeating the countercurrent distribution of this fraction. For the same reasons, constituent C also contained small amounts of active impurities and inactive aglycone.

Constituent A was then easily crystallized as the hydrochloride. One g of the base was dissolved in 10 ml of a dioxane-water mixture (80 : 20 by volume) and the solution acidifed to pH 3—4 by addition of hydrochloric acid. Five volumes of anhydrous dioxane were then slowly added to the filtered solution, and rubidomycin hydrochloride crystallized out in beautiful orange-red needles, as shown in Fig. 2.

The homogeneity of the hydrochloride obtained was checked by counter-current distribution. For this purpose, 10 mg of the hydrochloride were made to undergo

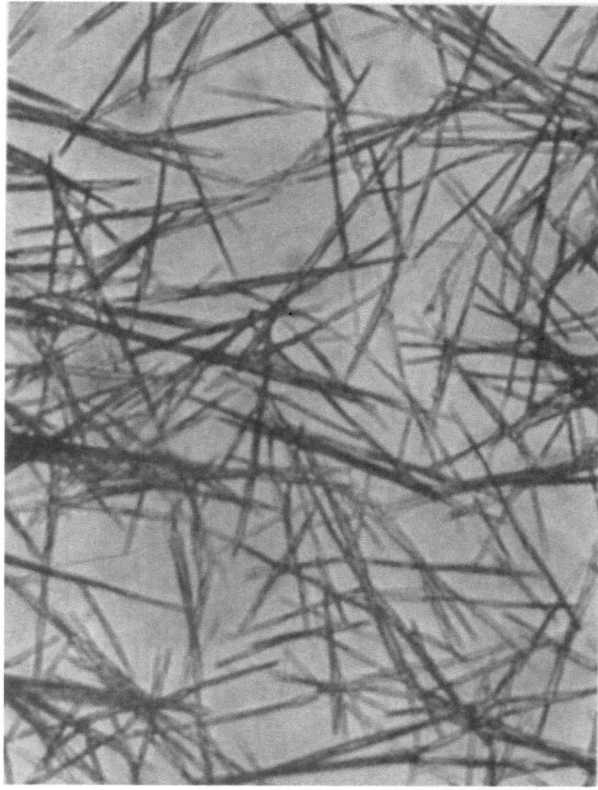

Fig. 2. Microphotograph of crystalline hydrochloride (magnification X 125)

60 transfers by distribution in a Craig apparatus consisting of 60 cells each cf 20 ml capacity, using the separate phases of the butanol-M/3 phosphate buffer solution system at pH 7.4 (1 : 1 by volume). The experimental distribution curve, determined

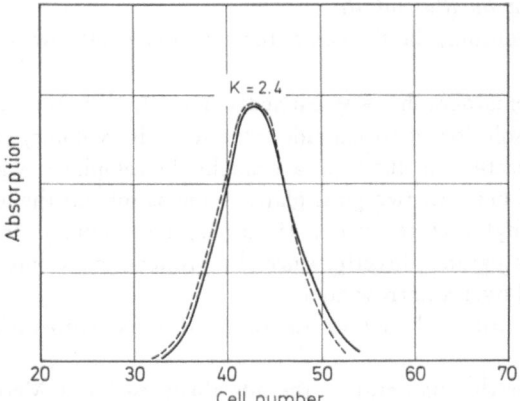

Fig. 3. Counter-current distribution of pure hydrochloride. ——— Experimental curve; - - - - - Theoretical curve

in the manner described above and shown in Fig. 3, coincided exactly with the calculated theoretical curve for K = 2.4 (peak corresponding to cells 42 and 43).

Constituents B and C could not be crystallized as salts because of their low stability in acid conditions.

3. Analytical Methods

The first attempts at isolating rubidomycin were rendered extremely difficult by the absence of any method of quantitative analysis. They were initially carried out by semi-quantitative assessment of the antitumor activity *in vivo* on mouse sarcoma 180. Then they were based on the *in vitro* measurement of the bacteriostatic activity of the broth and its various extracts on *Klebsiella pneumoniae*. Some degree of correlation was in fact found between the results obtained by the two methods.

The bacteriological method, however, was of very relative value, since, as well as rubidomycin, there are other substances in the broth possessing antibacterial activity, in particular the two substances B and C mentioned above. These substances are detectable as a group by the classical methods of dilution in a liquid medium or diffusion in agar.

These secondary constituents, which possess similar properties, were considered to be of less interest than rubidomycin.

The quantitative determination of rubidomycin was therefore possible only by the use of paper or thin layer chromatography, by which it could be separated from the other constituents. After separation, rubidomycin could be determined by colorimetric or microbiological methods.

Linear and circular paper chromatography and thin layer chromatography have both been used and both have been equally successful.

1. Linear chromatography was carried out on Arches No. 302 paper, impregnated with M/15 phosphate buffer at pH 4.5 using downward development. Development was effected by means of butanol saturated with water without previous equilibration, detection by means of the bioautographic method on plates of agar nutrient seeded with *Sarcina lutea* or *Bacillus subtilis*.

Under these conditions, the R_F values for the three constituents A, B, and C were 0.3—0.7, and 0.9.

2. Circular chromatography was carried out on Arches No. 302 paper, impregnated with a 20⁰/o solution of formamide in acetone (by volume) and then dried for an hour at a temperature of 40° C in a draught. Development was effected by the standard procedure, between two glass plates, using as solvent the upper phase of the system butanol—ethyl acetate—water (5 : 5 : 2.5 by volume). It was possible to examine the chromatograms directly, since the products were intensely colored and the spots distributed over a narrow zone.

Under these conditions, the R_F values for the three constituents A, B, and C were 0.1—0.3, and 0.9.

3. For thin layer chromatography the adsorbent used was Merck Kieselgel H in layers of 0.3 mm thickness. Two different methods were utilized:

I. The adsorbent was impregnated with M/15 phosphate buffer at pH 5, and the developing solvent was the upper phase of the system butanol—ethyl acetate—water (5 : 5 : 2.5 by volume).

II. The adsorbent was not impregnated, and the developing solvent was the upper phase of the system butanol—acetic acid—water (4 : 1 : 5 by volume).

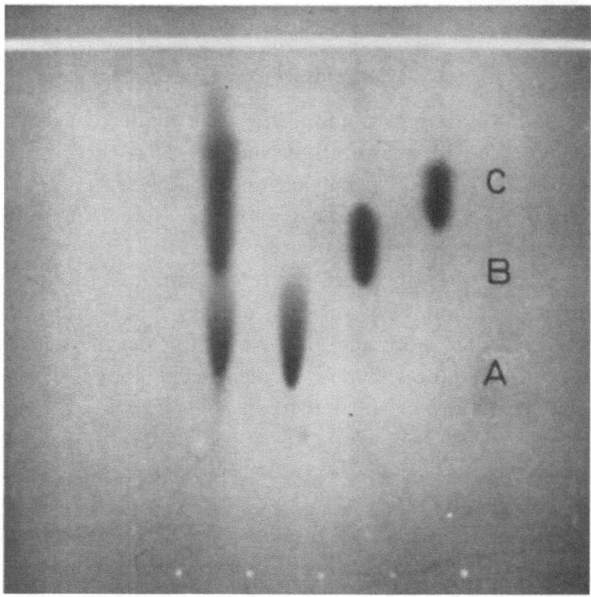

Fig. 4. Thin layer chromatography of rubidomycin (constituent A) and related substances (constituents B and C)

As the constituents were colored and collected into a number of small areas, it was possible here also to examine them directly.

Fig. 4 shows the chromatogram as prepared by method II.

Thin layer chromatography is particularly suitable for the quantitative determination of the constituents of a mixture. The colored areas can be scraped from the plate and eluted, then the individual constituents can be determined by spectrophotometry at 490 nm, comparing with standards chromatographed in parallel.

4. Physicochemical Properties

a) Solubility

Rubidomycin hydrochloride is readily soluble in water and physiological saline (solubility greater than 200 mg/ml), and also in methanol. It is slightly soluble in ethanol, and only very slightly soluble in acetone and chloroform. The base is soluble in chloroform, slightly soluble in alcohols, and practically insoluble in water.

b) Stability

In the solid state, the hydrochloride is very stable. It can be kept for several years at a temperature of 20—25° C without any loss of activity.

As shown by *in vitro* experiments (the agar diffusion technique of CHABBERT and VIAL [53]) and *in vivo* in mice with sarcoma 180, the aqueous solution of the hydrochloride does not lose any of its activity if kept for three weeks at 0° C or at 37° C.

As the lyophilized hydrochloride for clinical use, rubidomycin showed satisfactory stability in artificial aging tests, in which it was kept for 4, 8, and 15 days at 70° C or for six months at 45° C. In fact, it retained 85% of its initial microbiological activity after being kept for 15 days at 70° C, and 94% after being kept for six months at 45° C (Table 2).

Table 2. *Stability of lyophilized hydrochloride*

Property	Before heating	After heating						
		at 70° C			at 45° C			
		4 days	8 days	15 days	1 month	2 months	3 months	6 months
Appearance		spongy mass						
Color		orange-red						
Microbiological activity expressed as base (theoretical value = 10 mg)	9.8 mg	9.2 mg	8.6 mg	8.3 mg	10.5 mg	9.9 mg	9.3 mg	9.2 mg

The injectable solution, obtained by dissolving a quantity of lyophilized hydrochloride, corresponding to 10 mg base, in 2 ml of sterile distilled water, is very stable after being kept at room temperature (20—25° C) for 22 hours (Table 3).

Table 3. *Stability of aqueous solution of lyophilized hydrochloride*

Property	On prepa-ration	Storage at 20—25° C				
		2 hours	4 hours	5 hours	15 hours	22 hours
Appearance		clear to very slight cloudiness				
Color				orange-red		
Microbiological activity expressed as base (theoretical value = 10 mg)	9.45 mg	10.15 mg	9.2 mg	9.0 mg	9.45 mg	10.0 mg

c) Properties

The composition of the base corresponds to the formula $C_{27}H_{29}O_{10}N$. This base is strong enough to form a well-defined hydrochloride giving an aqueous solution of pH 5—6. The hydrochloride is strongly dextrorotatory, and its ultraviolet absorption spectrum has peaks that make it very easy to identify (Table 4).

Table 4. *Physicochemical properties of base and hydrochloride*

Property	Base		Chlorhydrate	
Empirical formula	$C_{27}H_{29}O_{10}N$		$C_{27}H_{29}O_{10}N, HCl, H_2O$	
Analytical results (%)	Calc.	Found	Calc.	Found
C	61.47	59.8	55.72	55.35
H	5.55	5.85	5.54	5.45
O	30.33	29.45	30.24	30.05
N	2.65	2.3	2.41	2.55
Cl			6.08	6.05
H_2O			3.1	2.95
Molecular weight	527.54		582.02	
$[\alpha]_D^{23}$ (c = 0.2, ethanol + 1 g/l HCl 12 N)			$+222° \pm 5°$	
Ultraviolet and visible absorption spectra [a]				
Wave lengths of absorption peaks (nm)	$E_{1\,cm}^{1\%}$		$E_{1\,cm}^{1\%}$	
234	621		597	
252	426		395	
290	135		136	
475—480	230		196	
490—495	230		196	
530	127		104	
Fluorescence spectrum [a]				
Wave length of the excitation peak			470 nm	
Wave length of the fluorescence peaks			550 and 580 nm	

[a] Determined on solutions in ethanol.

The ultraviolet and visible absorption spectra of the hydrochloride is shown in Fig. 5, the fluorescence spectrum in Fig. 6, and the infra-red absorption spectrum in Fig. 7.

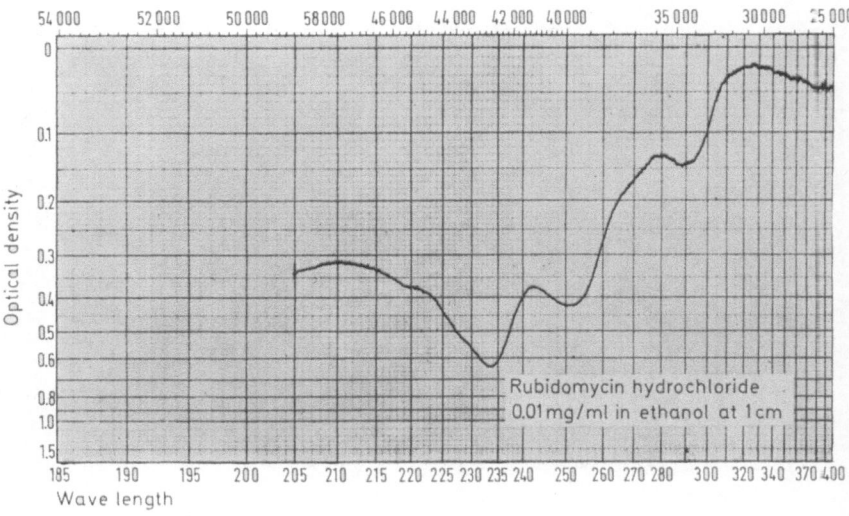

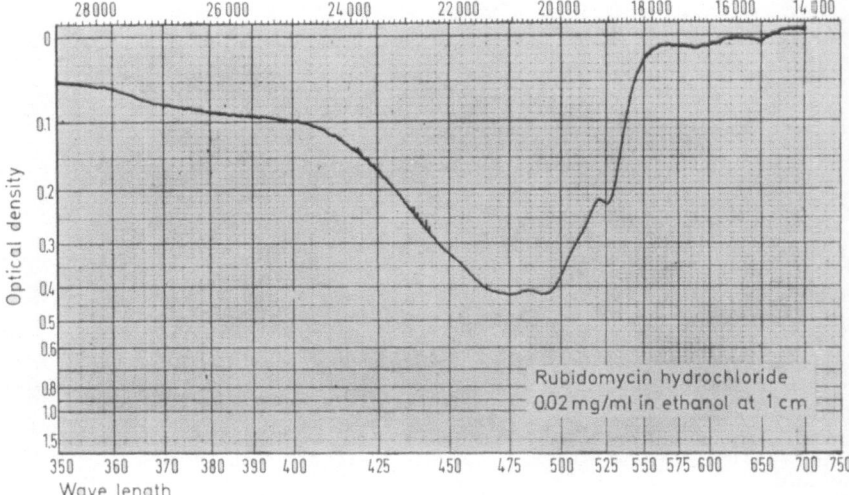

Fig. 5

Fig. 5. Ultraviolet and visible absorption spectra of the hydrochloride dissolved in ethanol

Fig. 6. Fluorescence spectrum of the hydrochloride dissolved in ethanol

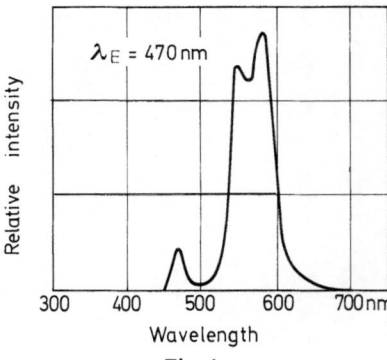

Fig. 6

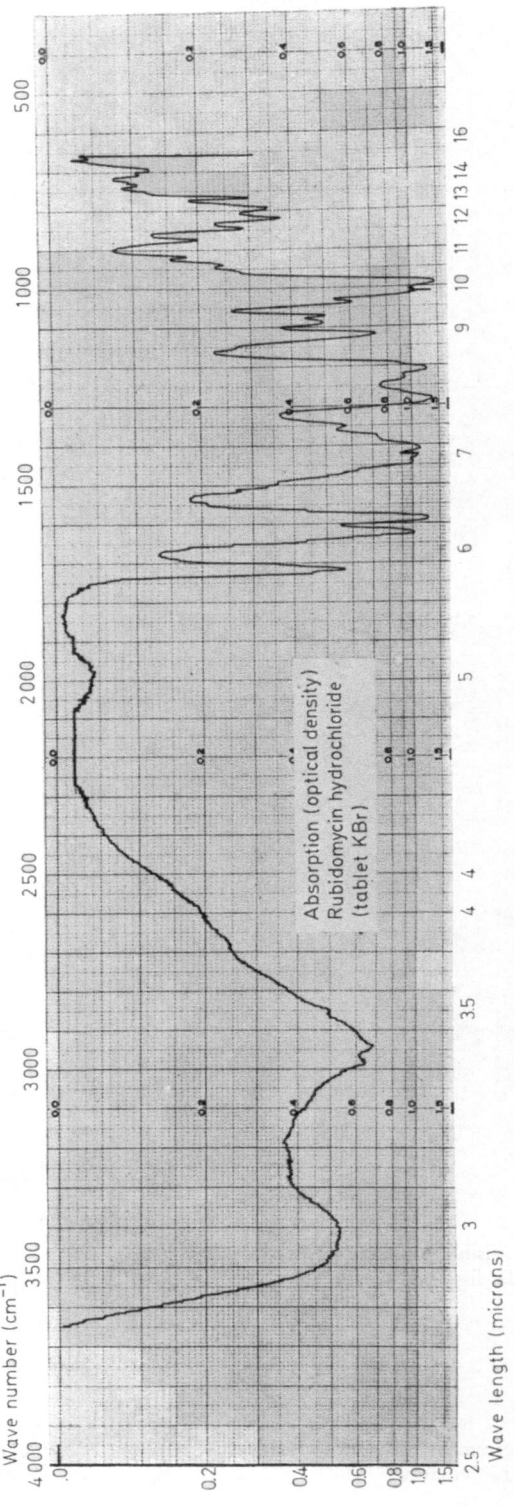

Fig. 7. Infra-red absorption spectrum of the hydrochloride

To facilitate the identification of rubidomycin, the chromophore was isolated by a method used for some anthracyclines.

Acid hydrolysis of rubidomycin with normal sulfuric acid for 20 minutes on a boiling water bath gave the aglycone—which was isolated in crystalline form—and an amino sugar. The most important properties of the aglycone are shown in Table 5.

Table 5. *Physicochemical properties of aglycone*

Property	Aglycone	
Appearance	crystalline powder	
Color	Orange-red	
Melting point	225—230°	
Empirical formula	$C_{21}H_{18}O_8$	
Analytical results (%)	Calc.	Found
C	63.31	63.4
H	4.55	4.7
O	32.13	30.4
Ultraviolet and visible absorption spectra (ethanol)		
Wave lengths of the absorption peaks (nm)	$E_{1\,cm}^{1\%}$	
234	835	
252	570	
289	167	
475	272	
490	272	
530	145	

Acid hydrolysis of constituents B and C gave the same aglycone as rubidomycin.

d) Comparison of Rubidomycin and Rubomycin C on the Basis of their Physicochemical Properties

Rubomycin has recently been described [38, 39, 163], as a mixture of the constituents A, B_0, B_1, and C. The workers concerned considered that the last of these constituents was identical with rubidomycin.

The identity of the two substances, which was based on the analytical data, has been confirmed by direct physical tests carried out on an authentic sample of rubomycin C kindly supplied by Professor GAUSE.

Thin layer chromatography was carried out with Merck Kieselgel H, either as it was or impregnated with M/15 phosphate buffer at pH 4.8. One of the following systems was used for development which was carried out at 25° C:

I. The upper phase of the system butanol-ethyl acetate-M/15 phosphate buffer at pH 4.8 (5 : 5 : 2.5 by volume) with impregnated layer.

II. The upper phase of the system butanol-M/15 phosphate buffer at pH 4.8 (5 : 2.5 by volume) with impregnated layer.

III. The upper phase of the system butanol-acetic acid-water (4 : 1 : 5 by volume) with non-impregnated layer.

The comparison data are found in Table 6.

Table 6. *Physicochemical properties of rubidomycin hydrochloride and rubomycin C hydrochloride*

Property	Rubidomycin hydrochloride	Rubomycin C hydrochloride
Appearance	Microcrystalline powder	Microcrystalline powder
Color	orange-red	orange-red
Analytical results (%)		
C	55.35	54.70
H	5.45	5.80
O	30.05	
N	2.55	2.55
Cl	6.05	6.10
H_2O	2.95	

Ultraviolet and visible absorption spectra (ethanol)				
Wave lengths of the absorption peaks (nm)	λ max.	$E_{1cm}^{1\%}$	λ max.	$E_{1cm}^{1\%}$
	234	597	234	570
	252	395	252	405
	290	136	290	135
	475—480	196	478	188
	490—495	196	495	188
	530	104	530	98

Infra-red absorption spectrum	No significant differences	
Thin layer chromatography		
R_F { System I	0.1—0.2	0.1—0.2
System II	0.1—0.2	0.1—0.2
System III	0.45	0.45

Under all the above conditions, rubidomycin and rubomycin C had identical R_F values, and the mixture of the two could not be dissociated. The published data for rubomycin C [38, 39] and this direct comparison of the physicochemical properties of rubidomycin and rubomycin C lead to the conclusion that the two substances are identical.

5. Tritiated Rubidomycin

It was found necessary to use radioactive rubidomycin to study its distribution and metabolism in animals.

a) Preparation

Tritiation was carried out by Wilzbach's method at the Commissariat à l'Energie Atomique (Saclay)—500 mg of pure rubidomycin hydrochloride was left in contact with 50 curies of tritium for three weeks.

The product of the reaction was dissolved in methanol. Thin layer chromatography of an aliquot, followed by both colorimetric and radiographic studies, showed that the product contained tritiated rubidomycin but that there were also tritiated impurities.

The pure hydrochloride was obtained by the following procedures:

1. Ether precipitation. The methanol solution was poured into 20 volumes of ether and the precipitate obtained filtered, washed, and dried for one hour at 40° C under vacuum (0.5 torr).

2. Crystallization. The hydrochloride obtained as described above was dissolved in water and dioxane was added in two portions, one before filtration and one slowly after filtration, so as to cause crystallization. The crystals were allowed to stand for two hours, then were filtered and washed, first with a mixture of dioxane and water (97.5 : 2.5 by volume), then with anhydrous dioxane. Finally, they were dried for 16 hours at 50° C under vacuum (0.5 torr).

A yield of 90 mg of tritiated rubidomycin hydrochloride was obtained.

b) Analytical Results

Various determinations were carried out on the tritiated rubidomycin hydrochloride:

1. The biological activity, determined by turbidimetry, was 1010 μg/mg, the theoretical activity of pure rubidomycin hydrochloride being 935 μg/mg.

2. The ultraviolet and visible absorption spectra were normal.

3. The radioactivity was 125 μCi/mg.

4. Thin layer chromatography was carried out as described under "Analytical Methods".

Colorimetric examination of the chromatograms obtained by systems I and II with a Chromoscan showed that the two products were of equal purity. Radiographic examination showed the presence of one radioactive substance only. The radiochromatogram obtained with system II is shown in Fig. 8.

5. In order to check the tritium distribution, a small amount of product was hydrolysed so that the radioactivity of the aglycone and the amino sugar could be determined separately. For this purpose, a trial sample was dissolved in normal

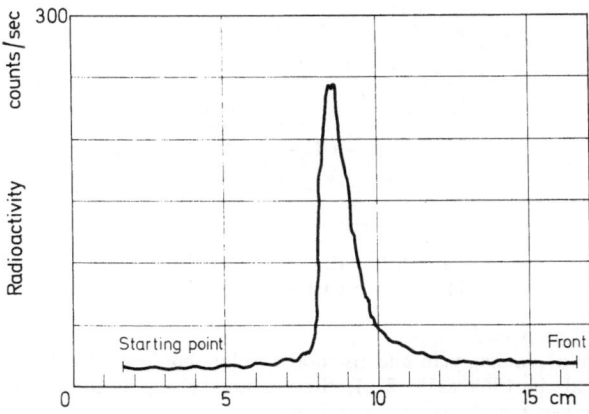

Fig. 8. Radiochromatogram of tritiated rubidomycin hydrochloride (Berthold scanner)

sulfuric acid and the solution refluxed for two hours. After cooling, the solution was extracted with ethyl acetate to separate the aglycone, which goes into the organic layer, from the amino sugar, which remains in the aqueous phase, and the radioactivity of the two solutions was measured after appropriate treatment.

$$\frac{\text{Radioactivity of aglycone}}{\text{Radioactivity of amino sugar}} = 2.57$$

$$\frac{\text{Number of hydrogen atoms in aglycone}}{\text{Number of hydrogen atoms in amino sugar}} = 1.42 \,.$$

This result shows that a slight difference exists in the distribution of the tritium between the two parts of the molecule.

As prepared by us, rubidomycin is stable to tritium, and it is easy to obtain a good yield of tritiated rubidomycin by the Wilzbach process.

Chapter 2

Structure

Daunomycin hydrochloride [1] (m.p. 188—190° decomp, $[\alpha]_D = +253°$ (methanol)) has an absorption spectrum in the ultraviolet and visible regions very similar to those of the 1,4,5-trihydroxyanthraquinones. It shows, in methanol solution, the following absorption peaks: $E_{1cm}^{1\%} = 665$, $\lambda = 234$ mμ; $E_{1cm}^{1\%} = 462$, $\lambda = 252$ mμ; $E_{1cm}^{1\%} = 153$, $\lambda = 290$ mμ; $E_{1cm}^{1\%} = 214$, $\lambda = 480$ mμ; $E_{1cm}^{1\%} = 218$, $\lambda = 495$ mμ; $E_{1cm}^{1\%} = 112$, $\lambda = 532$ mμ. The indicator-like behaviour, the metal complexing properties and the colors developed in sulfuric acid and in piperidine solutions are also typical of the above mentioned class of compounds. The presence of a quinoid chromophore is also shown by the readily reversible reduction on treatment with reducing agents as well as by the polarographic behaviour [51, 52].

Presently available evidence [3, 4, 7, 254] points to the structures Ia or Ib for daunomycin. The antibiotic is a glycoside formed by a tetracyclic quinoid aglycone (daunomycinone) and an amino sugar (daunosamine). Some details of the structure of daunomycin remain to be elucidated [2], namely the choice between structures I a and I b and the absolute stereochemistry at C-7, C-9 and C-1'.

I a: $-R_1 = -H$, $-R_2 = -OCH_3$
I b: $-R_1 = -OCH_3$, $-R_2 = -H$

---

[1] See footnote [2] on page 5.

[2] The structure of daunomycin and its total absolute configuration have been established by ARCAMONE et al. in 1968 (235, 236); the methoxy group is located on carbon 4 and specifications are given for all six asymmetric centers, i. e. 7 (S), 9 (S), 1' (R), 3' (S), 4' (S), 5' (S).

Daunomycin is chemically related to the anthracyclines, such as the rhodomycins, cinerubins, pyrromycins and rutilantins, whose chemistry has been reviewed by Brockmann [40]. The numbering system adopted by this author is illustrated in I. Daunomycin is therefore 7-O-(2',3',6'-trideoxy-3'-amino-L-lyxohexopyranosyl)-7,8,9,10-tetrahydro-9-C-acetyl-6,7,9,11-tetrahydroxy-1-(or 4)-methoxy-5,12-tetracenequinone.

The aglycone and the carbohydrate moieties of daunomycin are easily obtained by mild acid hydrolysis of the antibiotic. Structure II a or II b has been assigned to daunomycinone [7]. Daunomycinone (m.p. 213—214°, $[\alpha]_D = +193°$ (dioxane), empirical formula $C_{21}H_{18}O_8$) shows the same electronic spectrum as the parent glycoside.

R$_1$ O OH
.COCH$_3$
D C B A 'OH
R$_2$ O OH OH

II a: −R$_1$ = −H, −R$_2$ = −OCH$_3$
II b: −R$_1$ = −OCH$_3$, −R$_2$ = −H

R$_1$ O OAc
.COCH$_3$
'OAc
R$_2$ O OAc OAc

III a: −R$_1$ = −H, −R$_2$ = −OCH$_3$
III b: −R$_1$ = −OCH$_3$, −R$_2$ = −H

The tetracyclic structure was proved by the formation of tetracene on zinc-dust distillation. In the infra-red, a ketone group is observed (1718 cm^{-1}), in addition to the band for the chelated quinone (1617 cm^{-1}). Derivatives such as a semicarbazone (m.p. 232—235° decomp) and a dinitrophenyl hydrazone are easily obtained. In addition to a methoxyl group (Zeisel determination), four hydroxyls are present as shown by the conversion of daunomycinone, on treatment with acetic anhydride and pyridine, to a tetraacetate (III a or III b) (m.p. 225°, $[\alpha]_D = −95.5°$ (CHCl$_3$), which shows phenolic (1776 cm^{-1}) and alcoholic (1740 cm^{-1}) acetate bands, but no hydroxyl absorption in the infra-red. All the oxygen atoms of daunomycinone were thus accounted for, while the arrangement of the five groups containing these atoms on the anthraquinone portion of the molecule was suggested by the formation of salicylic acid on alkaline fusion of daunomycinone (and of its trimethyl ether as well).

Treatment of daunomycinone with either acids or alkalis gives a bisanhydro derivative (IV a or IV b) (m.p. 325—330°, conjugated ketone absorption (1685 cm^{-1}) in the infra-red) which in turn is converted to a diacetate (m.p. 240—243°, one OCH$_3$, phenolic acetate absorption (1765 cm^{-1}) in the infra-red) on acetylation. This proves that two hydroxyl groups are attached to ring A of daunomycinone.

R$_1$ O OH
.COCH$_3$
R$_2$ O OH

IV a: −R$_1$ = −H, −R$_2$ = −OCH$_3$
IV b: −R$_1$ = −OCH$_3$, −R$_2$ = −H

R$_1$ O OH
.COCH$_3$
'OH
R$_2$ O OH

V a: −R$_1$ = −H, −R$_2$ = −OCH$_3$
V b: −R$_1$ = −OCH$_3$, −R$_2$ = −H

The formation of a deoxy derivative on hydrogenation of daunomycinone with Pd on BaSO$_4$ is in agreement with the presence of a benzylic hydroxyl group on ring A. 7-deoxydaunomycinone (V a or V b) (m.p. 229—231°, $[\alpha]_D = −91°$ (CHCl$_3$), one OCH$_3$) yielded a triacetate (m.p. 126—128°) on acetylation.

The existence of an acetyl side chain and its attachment to a carbon atom bearing a hydroxyl group is proved by the formation of acetaldehyde, isolated as the 2,4-dinitrophenylhydrazone, on sodium borohydride reduction of daunomycinone, followed by periodate oxidation. Oxidation of bisanhydro daunomycinone with potassium permanganate (Equation 1) produced 3-methoxyphthalic acid (from ring D) and 1,2,4-benzenetricarboxylic acid (trimellitic acid, from ring A), thus proving that the methoxyl group is attached to C-1 or C-4, and that the acetyl side chain (and therefore a hydroxyl group in II a or II b) is attached to C-9.

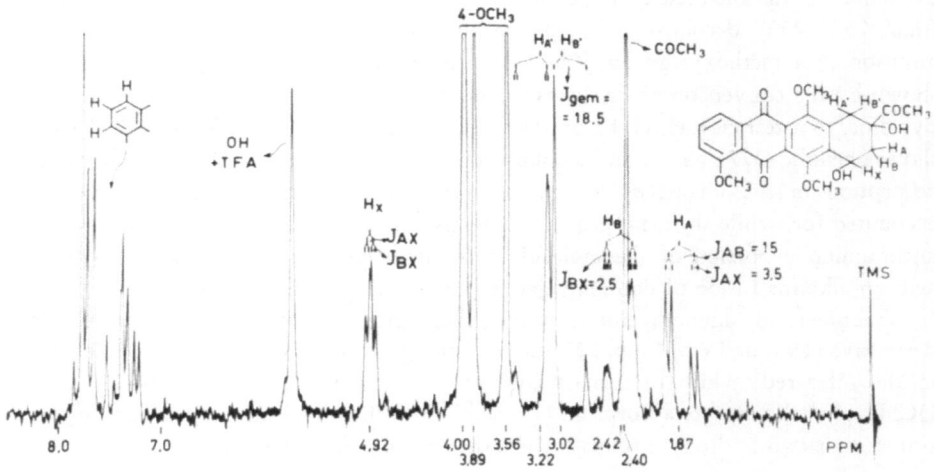

$$ \text{(1)} $$

The nuclear magnetic resonance spectra of daunomycinone and of its derivatives are in full agreement with the structures given above. The spectrum of daunomycinone trimethyl ether (Fig. 9) was particularly useful in the course of this work. This compound, prepared by refluxing daunomycinone with dimethyl sulfate in acetone in the presence of potassium carbonate, had a m.p. of 193°, four OCH_3 groups, $[\alpha]_D = +181°$ (dioxane), and a hydroxyl band at 3350 cm^{-1} in the infra-red.

Fig. 9. Nuclear magnetic resonance spectrum of daunomycinone trimethyl ether in CDCl$_3$. Varian A 60 spectrometer; chemical shifts are in p.p.m. (δ), relative to tetramethylsilane as internal standard

The n.m.r. spectrum shows two aromatic OCH_3 groups at 4.00 δ (p.p.m.), one aromatic OCH_3 group at 3.89 δ, an aliphatic OCH_3 group at 3.56 δ, while the acetyl group gives a signal at 2.40 δ. The C-9 free hydroxyl (singlet, 5.02 δ) is clearly recognized by the upfield shift with dilution and downfield shift with acid. The four lines signal at 4.92 δ (1 H), showing that the C-7 benzylic proton is the X part of an ABX spectrum, the AB part of which consists of two pairs of symmetric doublets centered approximately at 1.87 (1 H) and 2.42 δ (1 H); a first-order analysis gives

$J_{AB} = 15 \pm 0.2$; $J_{AX} = 3.5 \pm 0.2$; $J_{BX} = 2.5 \pm 0.2$ c.p.s. The magnitude of the J_{A3} shows geminal coupling, and the shifts of H_A and H_B are in agreement with a methylene β to the aromatic system. Two doublets (2 H, $J = 18.5 \pm 0.2$ c.p.s.) centered approximately at 3.02 and 3.22 δ indicate the two geminal benzylic protons at C-10, $H_{A'}$ and $H_{B'}$, without vicinal hydrogens. The three aromatic protons appear as a complex multiplet at 7—8 δ (3 H, ABC pattern).

Daunosamine, the amino sugar moiety of daunomycin, has been isolated as the hydrochloride (m.p. 168° decomp. $[\alpha]_D = -54.5°$ (H_2O) at equilibrium). It has been shown to be 2,3,6-trideoxy-3-amino-L-lyxohexose, corresponding to the structure and the stereochemistry illustrated in VI [5].

CH₃ … H, OH … HO NH₃⁺ H … Cl⁻

VI

CH₃ … H, OAc … AcO NHAc H

VII

Daunosamine is converted to the crystalline mixture of the anomeric triacetates VII (m.p. 168—170°, $[\alpha]_D = -71°$ (acetone)) on treatment with acetic anhydride and pyridine. Treatment of VI with 0.3 N methanolic HCl produces α-methyl-daunosaminide hydrochloride VIII (m.p. 188—190°, $[\alpha]_D = -130°$ (H_2O)) together with a minor amount of a less levorotatory compound. Methyl glycoside VIII is converted on acetylation to the O,N-diacetate IX (m.p. 188—189°; $[\alpha]_D = -202°$ ($CHCl_3$)). These data are a correction of the previously reported m.p. and $[\alpha]_D$ values for compound IX [5].

CH₃ … OCH₃ … HO NH₃⁺ Cl⁻ H

VIII

CH₃ … OCH₃ … AcO NHAc H

IX

Sodium periodate oxidation of VI yielded malonic dialdehyde and acetaldehyde, thus indicating that the deoxy groups are attached to C-2 (or C-3) and C-6. Compound VIII reduced one mole of periodate, thus proving the presence of the pyranose ring and the attachment of the deoxy group at C-2. The attachment of the amino group at C-3, together with the absolute configuration at this center were established by the periodate oxidation of N-benzoyl daunosamine X (m.p. 154—156°, $[\alpha]_D = -107.5°$ (ethanol)) prepared from VI with benzoyl chloride and aqueous sodium bicarbonate. This reaction (Equation 2) yielded acetaldehyde, originating from C-5 and C-6, and a non-volatile aldehyde which, on oxidation with sodium hypoiodite, was converted to the known N-benzoyl-L-(+)-aspartic acid XI, identical in all respects with an authentic specimen. The absolute configuration at C-3 was thus shown to be (S), following the convention of CAHN et al. [50].

$$
\begin{array}{c}
\text{CHOH} \\
|\\
\text{CH}_2 \\
|\\
\text{H-C*-NH-COC}_6\text{H}_5 \\
|\\
\text{H-C-OH} \\
|\\
\text{C-H} \\
|\\
\text{CH}_3
\end{array}
\xrightarrow{\text{NaIO}_4}
\begin{array}{c}
\text{CHO} \\
|\\
\text{CH}_2 \\
|\\
\text{H-C*-NH-COC}_6\text{H}_5 \\
|\\
\text{CHO} \\
\\
\text{CHO} \\
|\\
\text{CH}_3
\end{array}
\xrightarrow{\text{NaIO}}
$$

X

$$
\xrightarrow{\text{NaIO}}
\begin{array}{c}
\text{COOH} \\
|\\
\text{CH}_2 \\
|\\
\text{H-C*-NH-COC}_6\text{H}_5 \\
|\\
\text{COOH}
\end{array}
$$

(2)

$$[\alpha]_D = +33° \ (\text{H}_2\text{O plus 2 equiv. of KOH})$$

XI

The absolute configuration at C-4 and C-5, and thus the stereochemistry of daunosamine and of its derivatives, were established on the basis of rotational data, n.m.r. spectra and conformational arguments [3]. The comparison of molecular rotation is widely used in carbohydrate chemistry. The molecular rotation of daunosamine hydrochloride (-100) was compared with that of the equilibrium mixtures of the anomeric forms of the eight stereoisomeric 2,6-dideoxyhexoses. The L-lyxo configuration was suggested because only 2-deoxy-L-fucose and L-digitoxose show molecular rotation values approaching that of VI (-91 and -68 respectively), the L-digitoxose configuration (L-ribo) being ruled out because of the spatial arrangement at C-3 which is (R) in the L-ribo configuration.

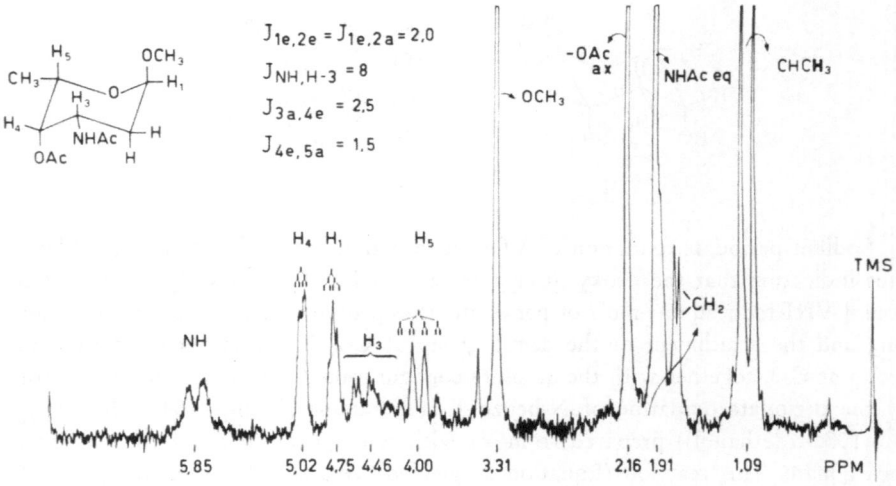

$$J_{1e,2e} = J_{1e,2a} = 2.0$$
$$J_{NH,H-3} = 8$$
$$J_{3a,4e} = 2.5$$
$$J_{4e,5a} = 1.5$$

Fig. 10. Nuclear magnetic resonance spectrum of α-methyldaunosaminide-N,O-diacetate in CDCl$_3$

[3] See footnote [2] page 18.

The n.m.r. spectra of IX (CDCl$_3$) (Fig. 10) show a triplet at 4.75 δ (H-1), the small splitting of which indicates the equatorial orientation of the anomeric proton, while the glycosidic OCH$_3$ group gives a signal at 3.31 δ, in agreement with an axial methoxyl group. H-3 appears as a broad signal at 4.46 δ, the line width of which, after exchange of amidic protons with deuterium on treatment with D$_2$O, measures 18—19 c.p.s. The H-4 signal (5.02 δ) is a quartet with splitting 1.5 and 2.5 c.p.s. H-5 is observed as a quartet (4.00 δ) which is further split to an octet ($J = 1$—1.5 c.p.s.) by H-4. The small coupling between H-5, H-4 (ca. 1 c.p.s.) and between H-4, H-3 (2.5 c.p.s.) excludes a diaxial orientation between the protons in question. The width of the H-3 signal suggests an axial orientation of this proton; the equatorial orientation of H-4, and the configuration (S) at C-4 can therefore be deduced. The stereochemistry at C-5 cannot be (R) (D-configuration) because in this case IX would exist in the more stable C-1 conformation which is not in agreement with the values of the coupling constants. The stereochemistry of daunosamine was thus established as 3(S), 4(S), 5(S), corresponding to the L-lyxo configuration. The configuration at C-1 (α-glycoside structure) was also suggested by the molecular rotation of VIII (-290), very near to that calculated for the unknown methyl 2-deoxy-α-L-fucoside (-273) [5], and is in agreement with the formation of a less levorotatory anomer in the preparation of VIII. The latter observation, daunosamine being a L-sugar, indicates VIII as the α-anomer.

These deductions appear to be correct because the structure and stereochemistry assigned to daunosamine are unequivocally supported by the synthesis of the natural Z sugar by Marsh et al. [166] and the stereospecific synthesis of derivatives of the D-enantiomorph worked out by Richardson [185]. Synthetic methyl 3-acetamido-2,3,6-trideoxy-α-D-lyxo-hexopyranoside (methyl α-D-daunosaminide-N,O-diacetate, m.p. 182—183°, $[\alpha]_D = +192$ (CHCl$_3$)) and 3-benzamido-2,3,6-trideoxy-D-lyxo-hexose (D-daunosamine-N-benzoate, m.p. 150.5—151.5, $[\alpha]_D = +110.5°$ (ethanol)) showed infra-red spectra almost identical with and superposable upon those of the corresponding naturally derived daunosamine derivatives IX and X.

The position of the linkage of the daunosamine moiety at C-7 of the aglycone is now established. The treatment of daunomycin with acetic anhydride and pyridine gives the fully acetylated compound (XII a or XII b).

XII a: $-R_1 = -H$, $-R_2 = -OCH_3$
XII b: $-R_1 = -OCH_3$, $-R_2 = -H$

In the n.m.r. spectrum of this compound, the signal of the C-7 proton is observed at about 5.0 δ (CDCl$_3$). If the benzylic hydroxyl group were free in the parent glycoside, the signal of the C-7 proton should be shifted in the range 6—6.5 δ on acetylation, as is observed in the n.m.r. spectrum of daunomycinone tetraacetate, in which the C-7 proton gives a signal at 6.35 δ (CDCl$_3$) [3]. The hydrogenation of daunomycin hydrochloride in methanol in the presence of Pd/BaSO$_4$ results in the hydrogenolysis of the benzylic glycosidic linkage yielding 7-deoxydaunomycinone (V a or V b) and daunosamine hydrochloride as the reaction products.

The rhodomycins A and B are typical anthracyclines, which show antitumor activity but quite unsuitable chemotherapeutic properties [6]. We may now compare the structure of rhodomycin B, XIII [45, 46], with I. This comparison shows that

XIII

great differences in the chemotherapeutic properties are related to minor variations in structural features. This observation suggests that further investigation of the chemistry of these antibiotics would probably result in the preparation of chemotherapeutic agents with more desirable antitumor and pharmacological properties than those that we have at present [4].

Chapter 3

Biological Activity

The choice of constituent A, i. e. rubidomycin, for use in chemotherapy [164, 165] was based on its greater stability, the fact that it could be obtained in the crystalline state [62] and, as found, in particular, with mice, its higher curative ratio; these factors more than outweighed its lower biological activity in absolute terms.

We investigated its cytostatic activity *in vitro,* its antitumor activity in animals, and its antibacterial activity *in vitro,* as well as its effects on the major systems of the animal organism.

In these tests, the antibiotic was used as the hydrochloride in solution either in physiological saline, when administered intravenously, or in distilled water, when administered orally or subcutaneously or intraperitoneally.

The doses used are always expressed in terms of the hydrochloride, which contains 93.5% of biologically active base.

1. Cytostatic Activity and Cell Changes

These tests were carried out on cells of four types: human KB and HeLa cells, mouse Ehrlich ascites cells, and plant cells.

---

[4] The synthesis of a tetracyclic analogue of daunomycinone and its conversion to the D glycoside are described by MARSH et al. [260].

a) KB and HeLa Cells

α) Cytostatic Activity

The cytostatic activity of rubidomycin was tested by four different methods: culture in tubes, culture on slides, culture in flasks, and diffusion in agar.

(a) Culture in Tubes (KB Cells)

KB cells (2×10^5 cells per ml) were grown at $37°$ C in tubes (18×150 mm) containing 1 ml of nutrient medium [1]. After 48 hours of growth, the medium was replaced by a fresh medium containing rubidomycin in various concentrations. After four days, the cultures were examined under a microscope with a low magnification. A concentration was considered to be cytotoxic when the cells were found to have degenerated.

Partial inhibition of cell growth was found at concentrations of rubidomycin of 2.5 µg/ml and above (degenerate and pyknotic cells), and it became complete at a concentration of 5.0 µg/ml.

(b) Culture on Slides (HeLa Cells)

HeLa strain cultures (human cervix carcinoma: 2×10^5 cells) were grown on slides and cover glasses, which were then placed in Petri dishes containing LE-medium (Behring), to which 15⁰/o of calf serum had been added. The preparations were incubated at $37°$ C for 24 hours.

On the day of the test, the nutrient medium was replaced by fresh material which, in the case of the cultures to be treated, contained rubidomycin in concentrations of 0.2, 1.0, 2.0 and 10 µg/ml. To determine the cytotoxicity of rubidomycin, the cover glasses were mounted on special slides after periods of 1, 3, 7, and 24 hours. The morphological changes in the structure of the tumor cells, disturbances in mitosis, and lethal damage to the tumor cells were recorded and photographed using combined film-phase-contrast microscopy. In addition, cells fixed by means of formalin were assessed after HE-staining.

In what follows, lethal damage is understood to mean the complete disintegration of the cell structure to form structureless cell debris. The proportion of lethally damaged cells within the complete culture, after rubidomycin in various concentrations had been allowed to act on it for various lengths of time, is shown in Fig. 11. The preparations were in all cases examined 24 hours after the beginning of the test. It was found that, for 50⁰/o of the tumor cells to be lethally damaged, a concentration of 2 µg/ml acting over a period of 24 hours was necessary. The extent of lethal damage was found to be a function essentially of concentration and time of exposure. At a concentration of 0.2 µg/ml, the proportion of lethally damaged cells increased only from 5.0⁰/o to 7.5⁰/o during the period of the test. Further tests with higher

---

[1] Nutrient medium:

Hanks salt solution   80 volumes
     Hydrolysate of lactalbumin   0.5⁰/o
     Difco yeast extract   0.1⁰/o
     Glucose   0.5⁰/o
Calf serum   20 volumes

(d) Diffusion in Agar (KB Cells)

The method used was based on that of CHABBERT and VIAL [53]. A layer of KB cells, 30 hours old, was covered with soft nutrient agar, on which discs impregnated with rubidomycin, at various concentrations, were placed. The agar was removed after 24 hours and replaced by a liquid nutrient medium. The areas in which cell growth had been inhibited were stained with fuchsin after 24 hours and their diameters measured in millimeters (absolute error: 2 mm approx.). The results obtained were plotted as a curve on semi-log paper, with the diameters as the ordinates and the concentrations as abscissae.

Under these conditions, the diameter is directly proportional to the concentration (Fig. 12).

β) Cell Changes

(a) KB Cells

KB cells were grown on slides in the same nutrient medium containing 0.1 or 0.5 µg/ml of rubidomycin. Each day, for four days, slides were taken, stained with May-Grünwald-Giemsa, and examined under the microscope, using an immersion lens.

As shown by the results given in Table 7, rubidomycin causes serious cytological changes at concentrations of 0.1 µg/ml and above.

Table 7. *Cytological action on KB cells*

Concentration of product (µg/ml)		Time of sampling		
	1 day	2 days	3 days	4 days
Controls	Small areas of epithelial cells	Areas of epithelial cells	Large areas of epithelial cells	Continuous sheet of epithelial cells
Rubidomycin 0.5	As for the controls	Cells fail to develop normally. Few areas of cells. Multinucleate cells	Degenerate cells: 70% Normal cells: 30%	Degenerate, pyknotic cells. Normal cells: 1%
0.1	As for the controls	As for the controls	As for the controls	Degenerate cells: 40% Normal cells (including some multinucleate cells): 60%

(b) HeLa Cells

Characteristic morphological changes were observed in the nucleoli of HeLa cells grown on slides after periods shorter than those for KB cells. At rubidomycin concentrations of 1 to 10 µg/ml, condensation and shrinkage of the nucleoli and fairly marked granulation of the nucleus were the main features. Seven hours after the

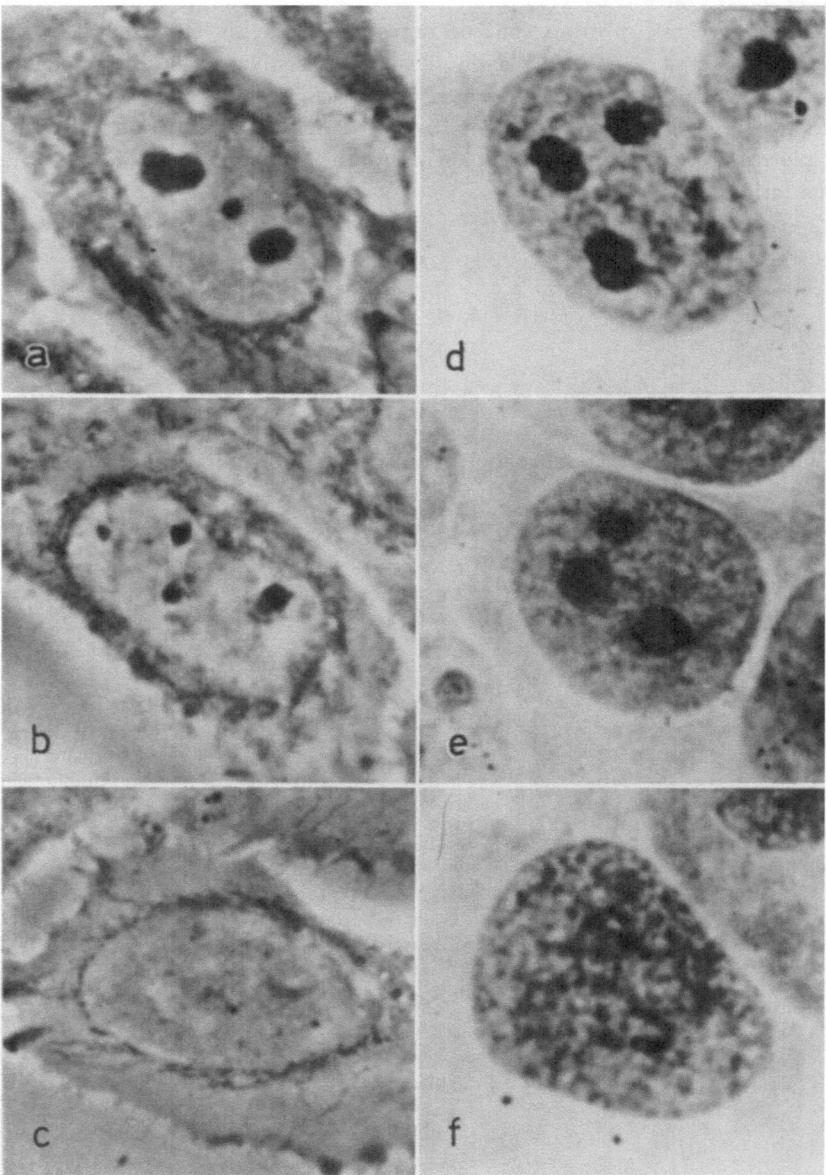

Fig. 13. The action on the structure of HeLa cells in relation to concentration after an exposure time of 7 hours. Left: phase contrast (magnification 500); Right: HE-staining (magnification 750). a control; b concentration 1 µg/ml; c concentration 10 µg/ml; d control; e concentration 0,2 µg/ml; f concentration 2 µg/ml

beginning of the experiment in which a concentration of 1 μg/ml was used, shrinkage of the nucleolus (Fig. 13 b) with condensation of residual structures was observed by phase-contrast microscopy. Finally, at a concentration of 10 μg/ml, only a few small granules remain in place of the original nucleolus (Fig. 13 c).

After a concentration of 0.2 μg/ml of rubidomycin had been allowed to act on the cells for seven hours, followed by HE-staining, the nucleolus appeared only to have become slightly opaque (Fig. 13 e). At a concentration of 2 μg/ml, the nucleolus could no longer be recognized with certainty in the markedly granulated nucleus (Fig. 13 f). The cytoplasm appeared unchanged over the whole range of concentrations. Both the thread-like structure of the mitochondria and their mobility were retained; the only noticeable change was that the formation of lipoid granules was inhibited.

In addition, the examination, under the electron microscope, of HeLa cells treated with the antibiotic [76, 203] also showed changes in the nucleoli, which reflect disturbances in the cell RNA.

b) Ehrlich Ascites Cells

α) Cytostatic Activity

This was studied by staining the cells with eosin.

Staining with Eosin

A cell suspension (2×10^6 cells per ml) of Ehrlich ascites tumor containing various concentrations of rubidomycin (3 to 200 μg/ml) was placed on the water-bath at 37° C for one hour. The cells were then stained with 1 ml of 0.1% eosin solution by the Schreck method [194]; the "dead" cells alone took the stain. The stained cells were counted so as to determine the proportion of "dead" cells.

A concentration of 100 μg/ml of rubidomycin gave 50% of "dead" cells (Table 8).

Table 8. *Action on Ehrlich cells, as shown by staining with eosin*

Concentration of rubidomycin (μg/ml)	Proportion of "dead" cells (as shown by eosin test) (%)	No. of mice dead after 40 days (Ehrlich ascites)
0 (controls)	10	20 d/20
500	90	0 d/10
100	50	0 d/10
50	26	0 d/10
25	14	5 d/10
12.5	20	10 d/10
6.25	14	10 d/10
3.12	10	10 d/10

In addition, ascites tumor cells, after treatment with the antibiotic, were injected intraperitoneally into G.S. mice [2] (10^6 cells per mouse).

Examination of the data given in Table 8 shows that there is lack of agreement between the results obtained *in vitro* and *in vivo*. In reality, the number of "dead"

---

[2] Heterozygotic albino mice bred in a closed colony.

cells must be greater than that shown by staining, since, for example, at a concentration of 50 µg/ml of rubidomycin, all the mice survived although the proportion of "dead" cells was only 26%.

β) Cell Changes

(a) Morphological Changes

The mice were given an intraperitoneal injection of 50 mg/kg of rubidomycin eight days after having been inoculated intraperitoneally with Ehrlich tumor cells (10^6 cells per mouse). The ascitic fluid was withdrawn each day on the first to the fourth days after the treatment, smeared on slides, stained with May-Grünwald-Giemsa, and examined under the microscope, using an immersion lens.

In spite of the high dose of rubidomycin given (about ten times as great as the therapeutic dose), cell changes were slow to appear. They appeared clearly only on the second day, and cell degeneration was practically complete only on the fourth day, while some malignant cells of unchanged appearance were invariably present.

(b) Effect on the Mitotic Index

Mice with Ehrlich ascites tumor were treated, six days after the injection of the cells, with 2.5 mg/kg of rubidomycin administered intraperitoneally. Under the same conditions, mice used as "colchicine controls" were given 50 µg/kg of colchicine intraperitoneally, and mice used as "absolute controls" were given physiological salt solution. The ascitic fluid was withdrawn 5, 24, and 48 hours after the treatment and

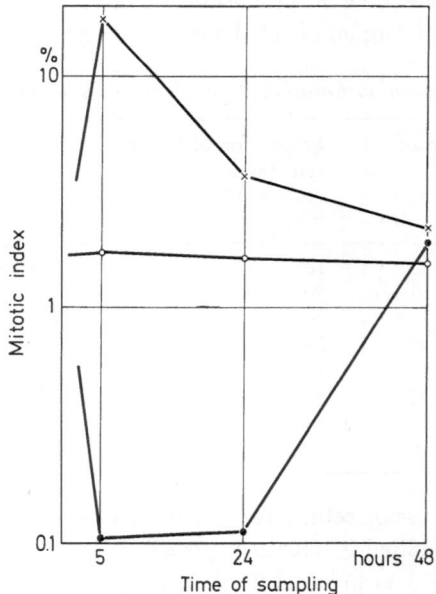

Fig. 14. Effect on the mitotic index (intraperitoneal treatment)
o——o Control; •——• Rubidomycin; ×——× Colchicine

centrifuged, and the cell sediment was smeared on a slide and fixed and stained by the Feulgen method so that the number of mitoses could be counted. The mitotic index (number of mitoses per 100 cells) was determined (Fig. 14) by the examination of more than 5000 cells in each sample.

Colchicine, chosen as a typical mitoclastic agent, increased the number of mitoses.

In contrast, rubidomycin may be regarded as a mitostatic agent [73, 162]; it considerably decreased the number of mitoses.

(c) Effect on Chromosomes

Rubidomycin was injected intraperitoneally at a dose of 5 mg/kg into mice inoculated with Ehrlich ascites tumor cells eight days previously. Samples of cells were removed 1, 4, and 16 hours after treatment. The chromosomes, after dispersion in a hypotonic solution and acid hydrolysis, were stained with crystal violet and examined under the microscope, using an immersion lens.

After four hours, the chromosomes of the treated cells were found to have changed, thickened, and often to have agglutinated, and some had even broken. Rubidomycin is therefore a mitostatic agent which changes the structure of the chromosomes.

c) Plant Cells

The action of rubidomycin on plant meristem cells (Allium test) has been studied by TRUHAUT and DEYSSON [220]. In the same way as with human and animal cells, the antibiotic was found to have mitodepressive chromatoclastic effects on the cells.

2. Antitumor Activity in Animals

The antitumor activity of rubidomycin on 18 tumors, of which four were of viral etiology, was investigated.

a) Tumors of Mice

Various types of tumors were used: implanted adenocarcinomas and a spontaneous mammary adenocarcinoma, Ehrlich ascites tumor, sarcomas, and leukoses.

In general, the doses of rubidomycin used were as follows:

(a) subcutaneously and intravenously   2.5   mg/kg
(b) intraperitoneally   1.25 mg/kg
(c) orally   12.5   mg/kg

Unless otherwise stated, treatment was begun on the day on which the tumor was implanted, and was continued for five days in the case of liquid and solid tumors, and for eight days in the case of leukoses.

The mice used as controls were treated with the solvent under the same experimental conditions.

Solid tumors were removed and weighed 12 days after implantation so that the percentage weight inhibition (W.I.) of the tumor could be calculated from:

$$W.I. = \frac{\text{Average weight of control tumors} - \text{average weight of treated tumors}}{\text{Average weight of control tumors}} \times 100 \; .$$

Inhibition was considered to be significant at values of W.I. greater than 55%.

The results of the tests are summarised in Table 13, the following symbols being used:

Marked activity (above 55%)   +
Slight activity (40—55%)   ±
No activity (less than 40%)   —

In the case of liquid tumors, i. e.. Ehrlich ascites tumor and the leukoses, the percentage prolongation of life (P.L.) as a result of the treatment was calculated from:

$$\text{P.L.} = \frac{\text{Average life of treated mice—average life of controls}}{\text{Average life of controls}} \times 100$$

Prolongation of life was considered to be significant at values of P.L. greater than 25%.

In the same way, the results of these tests are shown in Table 13, the following symbols being used:

Marked activity (greater than 25%)   +
Slight activity (20—25%)   ±
No activity (less than 20%)   —

In certain cases, account has been taken of the algebraic difference Δ between the average increase in weight of the controls and of the treated mice. This is an index of toxicity, and experience has shown that a high value of Δ means that any activity found experimentally is without significance; for example, in the case of sarcoma 180, Δ must be less than 3 g.

α) Carcinomas

(a) Mammary Adenocarcinomas

(α) Implanted Mammary Adenocarcinomas TM, TM 551, and Mn

Female mice, about one month old, were used. Adenocarcinoma TM or TM 551 was implanted subcutaneously by means of a trocar into RIII/Rho mice, and adeno-carcinoma Mn implanted in the same way into C_{57}Bl/Rho mice. In this case, the tumors were removed and weighed 15 days after implantation and the weight inhibition (W.I.) calculated.

As far as adenocarcinoma TM was concerned, rubidomycin showed marked inhibiting activity when administered subcutaneously (W.I. = 92%), moderate activity when administered intravenously (W.I. = 50%), and no activity when administered orally. The activity in the first case is of the same order as that of cyclophosphamide [3], triaziquone [4] and mercaptopurine [5].

As far as adenocarcinoma TM 551 was concerned, rubidomycin showed very marked activity when administered subcutaneously (W.I. = 100%), moderate activity when administered intraperitoneally (W.I. = 40%), and no activity when administered orally.

---

[3] Cyclophosphamide (Endoxan Asta-Werke, Cytoxan Mead-Johnson) is 1-bis(2-chloroethyl)amino-1-oxo-2-aza-5-oxaphosphoridin.

[4] Triaziquone (Trenimon Bayer) is 2,3,5-tris(ethyleneimino)-1,4-benzoquinone.

[5] Mercaptopurine (Purinethol Diamant) is 6-mercaptopurine.

As far as adenocarcinoma Mn was concerned, rubidomycin showed very marked activity when administered subcutaneously (W.I. = 96%).

(β) Spontaneous Mammary Adenocarcinoma RIII

Mammary carcinomas which appeared spontaneously in RIII/Rho mice were used. The average age of the mice was eight months, their actual ages varying from 6 to 11 months. Each batch, including both controls and treated mice, was made up of mice bearing non-necrotic tumors of homogeneous volume.

At the beginning of the test, each mouse was weighed and the surface area of the tumor measured in mm² by tracing it on to cardboard, which was then weighed. The animals were treated for 15 to 20 days. The surface area of the tumor was measured each week for one month. This was a severe test in which rubidomycin is only moderately active (slightly less than cyclophosphamide) when administered subcutaneously, and it appeared to keep the tumors in a stationary condition. It was without effect when given orally (Tables 9 and 13).

Table 9. *Subcutaneous and oral treatment of spontaneous mammary adenocarcinoma*

Product	Route	Daily dose (mg/kg)	Number of doses	Change in weight of tumors %	
				Stationary	Increase
Controls				20—10	80—90
Rubidomycin	s. c.	1.25	12	70	30
	oral	12.50	11	15	85

(b) Pulmonary Papillary Adenocarcinoma

This papillary tumor, which resembles an adenoma, was isolated in our laboratories in 1962 from a $C_{57}Bl/Rho$ mouse, and has been subsequently maintained by subcutaneous implantation.

Treatment, which was begun seven days after implantation, was continued for seven days. The tumors were removed and weighed on the 25th day.

The weight inhibition achieved by rubidomycin, when administered subcutaneously, was 59%. It was moderate (47%) intraperitoneally (Table 13).

β) Ehrlich Ascites Tumor

(a) Activity

G.S. mice, four weeks old, were inoculated intraperitoneally with 10^6 Ehrlich tumor cells in suspension in 0.5 ml of physiological saline – phosphate buffer at pH 7.2.

Rubidomycin was active when administered intraperitoneally (Table 10). Its activity was the same as that of triaziquone and, like that substance, it showed no activity when administered subcutaneously and orally.

Cyclophosphamide, methotrexate[6] and mercaptopurine have no effect on this tumor, whatever the route by which they are administered.

---

[6] Methotrexate (Amethopterin Lederle) is 4-amino-N¹⁰-methylpteroylglutamic acid.

Table 10. *Intraperitoneal treatment of Ehrlich tumor*

Daily dose of rubidomycin (over 5 days) injected intraperitoneally mg/kg	Prolongation of life (P.L.) (%)
1.25	60
0.62	30
0.31	30
0.15	6

(b) Attempt to Produce Cytological Resistance

Immediately after inoculation with the ascites tumor, the mice were given an intraperitoneal injection of 0.1 mg/kg of rubidomycin. In 13 serial passages, the tumor cells were injected intraperitoneally into fresh mice, which were treated, in their turn, in the same way. The cells were as sensitive to rubidomycin after the 13th passage as the original cells.

(c) Combined Treatment with "Anti-Ehrlich Cell"-Serum

Over a period of five days, 25 ml/kg of rabbit "anti-Ehrlich cell" serum was administered intraperitoneally together with 1.25 mg/kg of rubidomycin by the same route, after which rubidomycin was given alone for a further two days.

As shown in Table 11, this combined treatment gave slightly better results than those obtained with the antibiotic alone.

Table 11. *Combined intraperitoneal treatment with "anti-Ehrlich cell" serum and rubidomycin*

Product	No. of dead mice	
	24th day	42nd day
Controls	20 d/20	
"Anti-Ehrlich cell" serum	8 d/20	17 d/20
Rubidomycin	5 d/20	20 d/20
"Anti-Ehrlich cell" serum and rubidomycin	4 d/20	13 d/20

γ) Sarcomas
(a) Solid Sarcoma 180

Immediate treatment

(α) Activity

Tumor fragments of volume about 1 mm³ of this sarcoma were implanted subcutaneously into G.S. mice, about six weeks old, by means of a trocar.

Rubidomycin showed high activity when administered subcutaneously (Table 12), moderate activity (W.I. = 40%) when administered intravenously, and no activity when administered intraperitoneally and orally (Table 13).

When administered subcutaneously, rubidomycin showed the same activity as cyclophosphamide, methotrexate, triaziquone, and mercaptopurine.

Table 12. *Subcutaneous treatment of sarcoma 180*

Daily dose of rubidomycin (over 5 days) injected subcutaneously mg/kg	Weight inhibition (W.I.) ($^0/_0$)	$\varDelta$ (g)
5.000	81	3.4
2.500	83	0.1
1.250	79	−0.5
0.625	64	0.4
0.312	56	0.5

(β) Histological Examination of Tumors after subcutaneous Treatment

The tumors from treated mice, removed 12 days after implantation, were examined histologically. It was found that the tumors were composed of malignant cells which were very unusual because of their marked degeneration: they were swollen, rich in cytoplasm, with nuclei of varying size, sometimes enormous, deformed, and irregular, and with the chromatin often aggregated into enormous lumps.

(γ) Therapeutic Ratio

This ratio was calculated by a method resembling that of SKIPPER and SCHMIDT [205]. The weight inhibition obtained by means of the usual subcutaneous treatment was determined in mice in which sarcoma 180 had been implanted, and the toxicity, for the same route, found by tests on normal mice. The $90^0/_0$ curative dose (CD_{90}) and the $10^0/_0$ lethal dose (LD_{10}) were found graphically. The therapeutic ratio is given by $\dfrac{LD_{10}}{CD_{90}}$.

The therapeutic ratios found were as follows:

rubidomycin and methotrexate	10
cyclophosphamide	4
triaziquone	14

Delayed treatment

Instead of being started on the same day as the implantation, subcutaneous treatment was begun five days afterwards and was continued for five days. In this case, the weight inhibition was practically zero (W.I. $= 12^0/_0$, $\varDelta = 1$g).

(b) Sarcoma Induced by Benzpyrene

This tumor has been maintained in our laboratories for several years by subcutaneous implantation in $C_{57}Bl/Rho$ mice. Treatment was carried out and the results assessed in the same way as for sarcoma 180.

Rubidomycin was highly active when administered subcutaneously (W.I. $= 94^0/_0$), but inactive when administered by other routes (Table 13). When administered subcutaneously, it had the same activity as cyclophosphamide, triaziquone, and mercaptopurine.

(c) Reticulosarcoma

This spontaneous tumor was isolated in our laboratories in a $C_{57}Bl/Rho$ mouse, and has been maintained for four years by subcutaneous implantation. It is associated

Table 13. *Antitumor activity of rubidomycin in mice*
(Summary)

Type of tumor	Study material			Results	
	Strain of mice	Route of implantation	Route of administration of treatment (a)	Method of assessing results (b)	Activity (c)
Mammary adeno-carcinoma TM	RIII/Rho	s. c.	s. c. i. v. oral	W.I. W.I. W.I.	+ ± −
Mammary adeno-carcinoma TM 551	RIII/Rho	s. c.	s. c. i. p. oral	W.I. W.I. W.I.	+ ± −
Mammary adeno-carcinoma Mn	C$_{57}$Bl/Rho	s. c.	s. c.	W.I.	+
Mammary adeno-carcinoma	RIII/Rho	Spontaneous tumor	s. c. oral	W.I. W.I.	± −
Pulmonary papillary adenocarcinoma	C$_{57}$Bl/Rho	s. c.	s. c. i. p.	W.I. W.I.	+ ±
Ehrlich (ascites)	G.S.	i. p.	s. c. i. p. i. v. oral	P.L. P.L. P.L. P.L.	− + + −
Sarcoma 180 (solid)	G.S.	s. c.	s. c. i. p. i. v. oral	W.I. W.I. W.I. W.I.	+ − ± −
Sarcoma induced by benzpyrene	C$_{57}$Bl/Rho	s. c.	s. c. i. p. i. v. oral	W.I. W.I. W.I. W.I.	+ − − −
Reticulosarcoma	C$_{57}$Bl/Rho	s. c.	i. p. i. v.	W.I. and P.L. P.L.	+ +
Lymphoblastic leukemia AKR	AKR/Rho	i. p. intracerebral	i. p. i. v. oral i. p. oral	P.L. P.L. P.L. P.L. P.L.	+ + − ± −
Lymphoblastic leukemia X	C$_{57}$Bl/Rho	i. p.	i. p.	P.L.	+
Myeloid leukemia C 1498	C$_{57}$Bl/Rho	i. p.	s. c. i. p. oral	P.L. P.L. P.L.	− + −
Leukemia L 1210	B$_6$D$_2$F$_1$/Cum	i. p.	i. p.	P.L.	+
Leukosarcomatosis	C$_{57}$Bl/Rho	i. p.	i. p. i. v.	P.L. P.L.	+ +

(a) Treatment with the maximum tolerated dose not causing any significant loss of weight.
(b) W.I. — weight inhibition (%); P.L. — prolongation of life (%).
(c) Marked activity: +; slight activity: ±; no activity: −.

with hypertrophy of the axillary, inguinal, and submaxillary lymphnodes and with splenomegaly.

When administered intraperitoneally 19 days after implantation, rubidomycin gave a weight inhibition for the spleen of 55% and a prolongation of life of 25%, an activity comparable with that of methotrexate. Cyclophosphamide remains the drug with the highest activity. When administered intravenously to mice, rubicomycin gave a prolongation of life of 30%.

δ) Leukoses and Leukosarcomatosis

(a) Leukoses

The experimental protocol was identical for the four types of leukosis studied, the only changes being in the strain of mice used. Male or female mice, about one month old and weighing 18—20 g, were inoculated intraperitoneally with 10^6 leukaemic cells per mouse (except in the case of leukemia L 1210, where they were inoculated with 10^4 cells) contained in 0.5 ml of physiological saline, or intracerebrally with 10^3 cells in 0.05 ml. Treatment, which was begun on the day of the inoculation, was continued for eight days. The activity of the drug used was assessed from the increase in the prolongation of life (page 32).

(α) Lymphoblastic Leukemia AKR

AKR/Rho mice were inoculated with this leukemia.

In the case of intraperitoneal inoculation, rubidomycin showed appreciable activity when administered intravenously (P.L. = 35%), and much greater activity when administered intraperitoneally (P.L. = 100%). By this route its activity was equal to that of cyclophosphamide and greater than that of methotrexate (P.L. = 40%).

In the case of intracerebral inoculation, rubidomycin showed only moderate activity when administered intraperitoneally (P.L. = 24%).

It was inactive in all cases when administered orally (Table 13).

(β) Lymphoblastic Leukemia X

$C_{57}Bl$/Rho mice were inoculated with this leukemia.

When administered intraperitoneally, rubidomycin was highly active (P.L. = 160%) against this leukemia.

(γ) Myeloid Leukemia C 1498

Rubidomycin showed appreciable activity (P.L. = 62%) when administered intraperitoneally to mice inoculated with this leukemia, but was inactive when administered subcutaneously and orally (Table 13).

When administered intraperitoneally, methotrexate and mercaptopurine were inactive. Cyclophosphamide showed the same activity as rubidomycin.

(δ) Leukemia L 1210

In $B_6D_2F_1$/Cum mice, inoculated intraperitoneally with 10^4 cells, rubidomycin showed marked activity (P.L. greater than 150%) when administered intraperitoneally—appreciably greater than that shown by methotrexate and cyclophosphamide.

(b) Leukosarcomatosis

This malignant disease, which is sensitive to the action of asparaginase ("asparagine dependent" cells), appears spontaneously in $C_{57}Bl/Rho$ mice; it has been maintained in the laboratory for five years by intraperitoneal inoculation. Inoculated mice generally die after 12 days.

Rubidomycin was active in intraperitoneal doses of 0.5 mg/kg and in intravenous doses of 1.25 mg/kg, which gave a prolongation of life of 49% and 51% respectively.

When administered intraperitoneally, methotrexate was inactive, but cyclophosphamide had an activity four times as great as that of rubidomycin.

b) Tumors and Leukemia of Viral Etiology

α) Avian Sarcomas

(a) Rous Sarcoma

Wyandotte chicks, two days old, were inoculated subcutaneously in the wing web with 10^3 ID_{50} of Rous sarcoma virus (Bryan strain). Rubidomycin was administered subcutaneously at a daily dose of 0.5 mg/kg, treatment being begun three days after the inoculation and being continued for five consecutive days. Under these conditions, the antibiotic had no effect on the development of the tumors which, in the treated animals, appeared after the same length of time (seven days after the inoculation of the virus) and developed in the same way as in the controls.

Under the same experimental conditions, cyclophosphamide administered intraperitoneally at a daily dose of 150 mg/kg showed a marked effect on the appearance of the tumors, at the cost of high toxicity to the chicks, the effective dose being close to the LD_{50} of the drug. In contrast, even at the toxic dose of 0.8 mg/kg, rubidomycin administered intraperitoneally had no effect on the tumors caused by the Rous sarcoma virus.

(b) Sarcoma Caused by a Recently Isolated Virus

We have demonstrated the presence [183], in cultures of embryonic cells of the quail, of a latent virus which is cytopathogenic and tumorigenic for quails and chicks.

Investigations have been carried out to determine whether rubidomycin was active against the sarcoma produced in chicks by this virus.

Male Leghorn chicks, five days old, were inoculated subcutaneously in the wing web with 100 ID_{50} of this virus. Rubidomycin was administered intraperitoneally at a daily dose of 60 µg/kg for five days, treatment being begun on the same day as the inoculation.

As in the case of the Rous sarcoma, rubidomycin was inactive against this sarcoma.

β) Shope Fibroma

The Patuxent strain of the Shope fibroma virus, at dilutions of 10^{-1} to 10^{-7}, was injected intradermally into the flanks of adult male rabbits belonging to the Géant Blanc du Bouscat strain. Treatment with rubidomycin was effected by subcutaneous injection of a daily dose of 0.5 mg/kg for three successive days, beginning immediately after the injection of the virus.

Fig. 15 shows the development of the tumors in the treated animals and the controls, in terms of the maximum dilution at which a palpable fibroma is produced in at least half the rabbits at various periods after the injection of the virus.

These tests showed that rubidomycin had a significant inhibiting effect on the development of this tumor, comparable to that obtained with a dose of cyclophosphamide 100 times as large, administered intravenously.

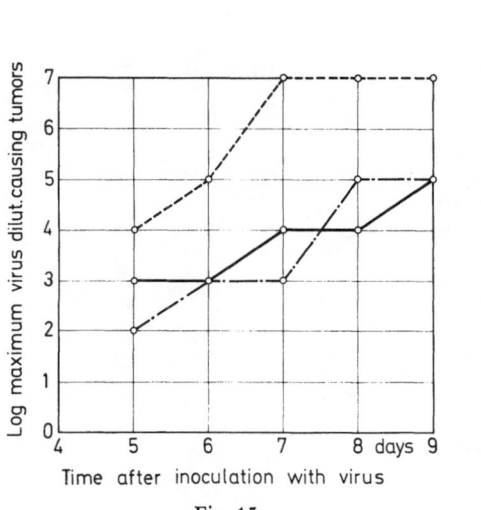 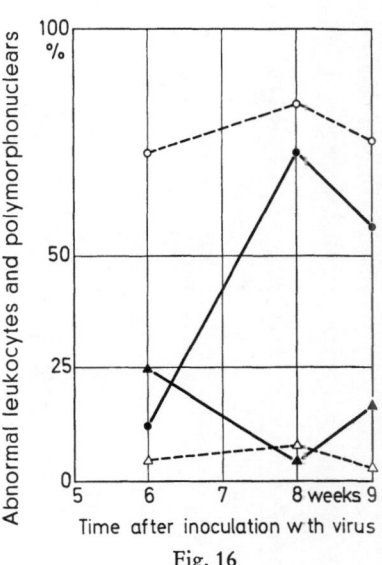

Fig. 15   Fig. 16

Fig. 15. Effect on the development of Shope fibroma (subcutaneous treatment)
o – – – o Control; o ——— o Rubidomycin; o —·— o Cyclophosphamide

Fig. 16. Rauscher leukemia (intraperitoneal treatment). Effect on hematological symptoms

Leukocytes abnormal { control   o – – – o
                      { rubidomycin   ● ——— ●
Polymorphonuclear { control   △ – – – △
           cells { rubidomycin   ▲ ——— ▲

γ) Rauscher Leukemia

Male BALB/c mice, between five and six weeks old, were inoculated intraperitoneally with the Rauscher leukemogenic virus. Treatment with rubidomycin, administered intraperitoneally at a daily dose of 0.5 mg/kg, was begun one week after the inoculation with the virus and was continued for only three consecutive days. Six weeks after inoculation with the virus, i. e., 32 days after the conclusion of the treatment, 20 controls and 20 treated mice were sacrificed and the average weight of their spleens determined. Values of 1950 mg and 537 mg were respectively found for the controls and the mice treated with rubidomycin as compared with 190 mg for uninoculated and untreated mice. Blood smears were made six, eight, and nine weeks after inoculation with the virus; the average percentages of abnormal leukocytes and polymorphonuclear cells found in the controls and the treated mice are shown in Fig. 16. It will be seen that there is a marked difference between the controls and the treated mice in the tests carried out six weeks after the inoculation, the percentage of

abnormal leukocytes being small and that of the polymorphonuclear cells being
normal in the case of the latter. The difference was considerably less in the tests
carried out eight weeks after inoculation of the virus, but was again significant nine
weeks after inoculation.

Figs 17 and 18 show blood smears made six weeks after the inoculation of BALB/c
mice with Rauscher virus; Fig. 17 control, Fig. 18 treated with rubidomycin.

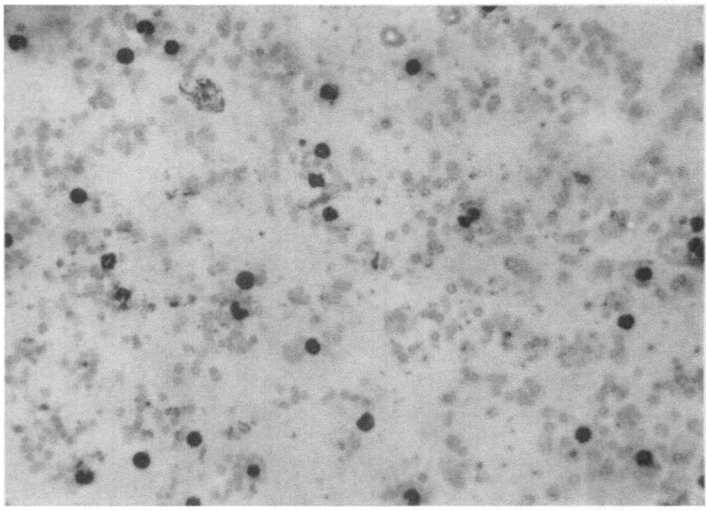

Fig. 17. Blood smear from BALB/c mouse inoculated with Rauscher leukemia virus and
untreated (control). Stained with May-Grünwald-Giemsa stain; magnification ×100

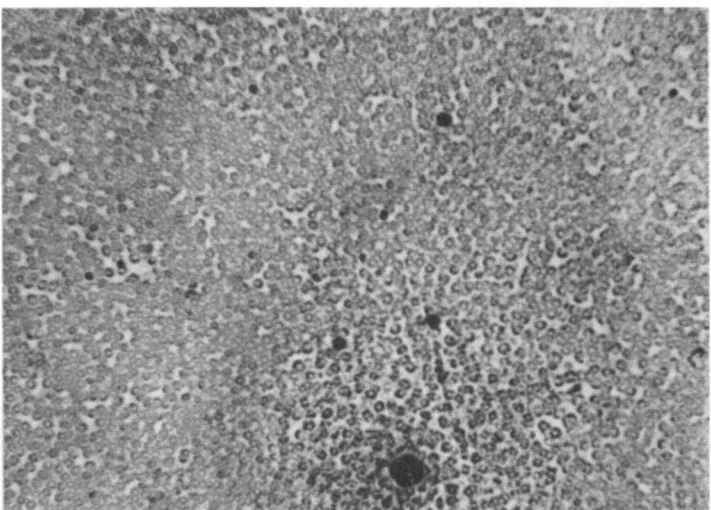

Fig. 18. Blood smear from BALB/c mouse inoculated with Rauscher leukemia virus and treated
by the intraperitoneal route. Stained with May-Grünwald-Giemsa stain; magnification ×100

As far as the controls are concerned (Fig. 17), a marked hyperlymphocytosis will be noted, together with considerable changes in the lymphocytes (dense nuclei with ragged edges and cytoplasm undergoing lysis). Severe anemia will be observed, together with anisocytosis and poikilocytosis of the erythrocytes, several of which contain Jolly bodies.

As far as the treated mice are concerned (Fig. 18), the smear is practically normal in appearance, although the erythrocytes show some hypochromatism.

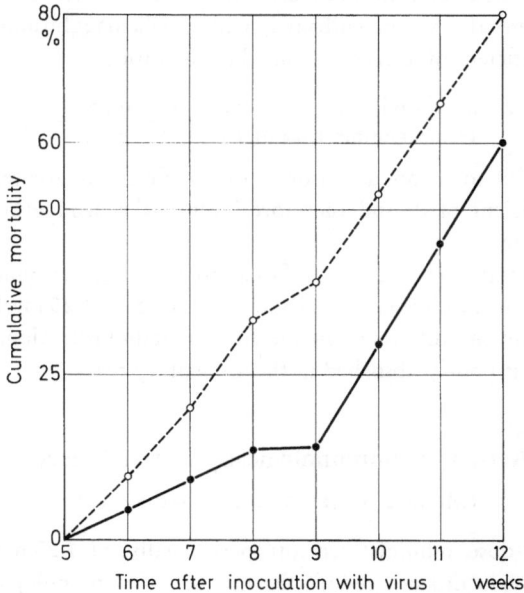

Fig. 19. Rauscher leukemia (intraperitoneal treatment). Effect on mortality
o ----- o Control;   •——• Rubidomycin

Fig. 19 shows the cumulative mortality curve during the 12 weeks following inoculation with the virus (20 treated mice and 20 controls). It is considerably slower in the case of the mice treated with rubidomycin.

Our results on the effect of rubidomycin on the leukemia caused by the Rauscher virus are not comparable with those obtained by the use of other drugs. In their case—with, for example, cyclophosphamide—treatment was continued for a much longer period (two to three weeks), and their action on splenomegaly and the percentage of atypical leukocytes was obtained only at the cost of high toxicity, as shown by loss of weight, marked leukopenia, and an appreciable mortality during the treatment. None of these toxic effects was observed in the tests with rubidomycin described above.

Thus, treatment with rubidomycin over a very short period had a marked effect on the mouse leukemia caused by the Rauscher virus, as shown by the blood picture and the mortality. It should be noted that, during a further series of tests, we found that the action of rubidomycin on Rauscher leukemia was somewhat variable.

3. Immunodepressive Activity

a) Effect on the Production of Antibodies

It is well known that certain antibiotics inhibit the production of antibodies [13]. In our tests, we used sheep erythrocytes as the antigen, each six-week old male RIII/Rho mouse being injected intravenously with 10^8 of them. Rubidomycin was administered intraperitoneally, at a dose of 1.25 mg/kg, 48 hours before the antigen, and then again one hour and 48 hours after it. The mice were bled 10 days after the injection of antigen, the serum collected, and the hemagglutinin titer determined. The inhibition coefficient of a drug is defined as the ratio:

$$\frac{\text{Hemagglutinin titer of serum of treated animals}}{\text{Hemagglutinin titer of serum of controls}}.$$

A drug is considered to be an inhibitor if its coefficient is greater than 1. Rubidomycin has a coefficient of 8, and therefore inhibits the production of heterohemagglutinins in this system.

SHORINE and SHAPOVALOVA [200] failed to find any immunodepressive action when rubomycin was administered orally in doses of 5 and 25 mg/kg. This difference can be explained by the difference in the route of administration, since it is known that this antibiotic is poorly absorbed in the alimentary tract.

b) Action on Immunologically Active Spleen Cells

(Method of JERNE and NORDIN [138])

At an effective dose, administered intraperitoneally, of 1.25 mg/kg, which is well tolerated, rubidomycin did not reduce the number of immunologically active spleen cells in mice. A reduction in the number of these cells was achieved only at the toxic dose, administered intraperitoneally, of 2.5 and 2.75 mg/kg.

c) Action on Grafts

Female AKR/Rho mice, about two months old, received grafts of skin from female $C_{57}Bl$/Rho mice of the same age (donors and recipients differ in H_2). The recipients were treated two hours before grafting, and then every day until the graft was rejected, with 0.625 mg/kg of rubidomycin administered intraperitoneally. On the eleventh day the dressing was removed, and the date on which the graft was rejected was recorded. Another batch of AKR/Rho mice received skin grafts in the same way, but was not treated, and therefore served as a control group.

Table 14. *Action on grafts (intraperitoneal treatment)*

Product (mg/kg) i. p.	Average duration of graft (days)
Controls	12.3 ± 0.95
Rubidomycin 0.625	14.8 ± 1.76

As shown in Table 14, rubidomycin slightly prolonged the period during which skin grafts were accepted by mice; this prolongation (2.5 days) was not significant. Under identical experimental conditions, azathioprine [7] had no effect.

d) Action on the Lymphoblastic Transformation of Lymphocytes
in vitro

Mixtures of lymphocytes from two dogs of different breeds were grown *in vitro* by the method of BACH and HIRSCHHORN [9] and BAIN et al. [10].

In the mixed cultures used as controls, 4% of the lymphocytes were converted into typical lymphoblasts. In the presence of 0.05 µg/ml of rubidomycin, no lymphoblastic transformation was observed, and more than 50% of the lymphocytes were destroyed. In the presence of 0.01 µg/ml, only 2% of the lymphocytes underwent lymphoblastic transformation and the remaining cells did not undergo any detectable change.

In a subsequent experiment, carried out on a mixture of lymphocytes from two other dogs, a lymphoblastic transformation of 8% of the cells in the mixed control cultures was observed, while in the mixed cultures containing 0.025 µg/ml of rubidomycin only 1% of the lymphocytes underwent this transformation. It should be noted, on the other hand, that even at a concentration of 0.1 µg/ml of rubidomycin this substance had no inhibiting effect on the lymphoblastic transformation of dog lymphocytes exposed to the action of phytohemagglutinin, in spite of the fact that the lymphocytes had been incubated for 24 hours in the presence of rubidomycin before the phytohemagglutinin had been added. A concentration of 0.5 µg/ml of rubidomycin was alone capable of inhibiting significantly the effect of phytohemagglutinin on lymphocytes. It should be remembered that COSTA and ASTALDI [59, 60] showed that, at a concentration of 0.2 µg/ml, the antibiotic inhibited the lymphoblastic transformation of human lymphocytes cultured *in vitro* in the presence of phytohemagglutinin.

Our observations showed, therefore, that at very low concentrations (less than those which have a harmful effect on the lymphocytes) rubidomycin is capable of inhibiting the immunological reactivity of these cells.

Under the same experimental conditions, amethopterin at a concentration of 0.1 µg/ml had no effect on the lymphoblastic transformation of the lymphocytes.

Other tests were carried out by the method of GEORGE and VAUGHAN [103] in order to determine whether rubidomycin could counteract the inhibition by tuberculin of the *in vitro* migration in capillary tubes of the macrophages of guinea-pigs sensitized by means of complete Freund adjuvant (BCG). It was found that rubidomycin, which was toxic to the macrophages at a minimum concentration of 0.2 µg/ml, did not counteract the inhibiting effect of tuberculin when it was present in concentrations in the range 0.01 to 0.1 µg/ml in the culture medium of the macrophages.

4. Antibacterial Activity

In addition to its antimitotic activity, rubidomycin possesses antibacterial properties [62, 178] which are worth noting although, because of its toxicity [139], its use in antibacterial therapy in man or animals can hardly be envisaged.

---

[7] Azathioprine (Imuran Burroughs Wellcome) is 6-[(1-methyl-4-nitroimidazol-5-yl)thio] purine.

By way of illustration, the minimum bacteriostatic concentration for certain bacteria taken as representative of the most important species generally used in the laboratory are shown in Table 15. As will be seen from the Table, rubidomycin is effective as an antibacterial agent mainly in the case of gram-positive bacteria.

Table 15. *Antibacterial activity*

Organism	Smallest concentration inhibiting growth (µg/ml)
Staphylococcus aureus, strain 209 P — ATCC 6538 P	4.3
Staphylococcus aureus, strain Oxford — ATCC 9144	4.2
Sarcina lutea — ATCC 9341	1.2
Streptococcus pyogenes hemolyticus, strain Dig 7 — Institut Pasteur	1.6
Diplococcus pneumoniae, strain Til — Institut Pasteur	0.9
Corynebacterium pseudodiphtericum	1.7
Lactobacillus casei — ATCC 7469	1.2
Bacillus subtilis — ATCC 6633	1.2
Bacillus cereus — ATCC 6630	2.2
Bacillus brevis — ATCC 8185	1.4
Bacillus megatherium — NRRL B 1125	1.3
Bacillus polymyxa — NCTC 4747	4.4
Mycobacterium species — ATCC 607	1.2
Mycobacterium phlei — Institut bactériologique de Lyon	5
Mycobacterium para-smegmatis — A 75 — Lausanne	2.2
Escherichia coli — ATCC 9637	500
Shigella dysenteriae — Shiga L — Institut Pasteur	2000
Salmonella typhimurium — Institut Pasteur	1000
Aerobacter aerogenes — ATCC 8308	1000
Klebsiella pneumoniae — ATCC 10.031	8.9
Pseudomonas aeruginosa, strain Bass — Institut Pasteur	1000
Brucella bronchiseptica — CN 387 — Wellcome Institut	62
Pasteurella multocida — A 125 — Institut Pasteur	1.8

5. General Pharmacological Properties

The general pharmacology of rubidomycin was studied mainly in dogs (the animals were in all cases anesthetized with pentobarbital at an intramuscular dose of 40 mg/kg and an intravenous dose of 10 mg/kg), but a few additional tests were carried out on guinea-pigs and the papillary muscle of the heart of the cat.

a) Effect on the Cardiovascular System

α) Effect on the Arterial Pressure

In dogs, at a dose of 5 mg/kg administered intravenously, rubidomycin had no effect on the arterial pressure (the period of observation extended over five hours after the administration of the product). The only effect noted was a temporary lowering of the pressure on injection (of about 20%), with a return to the original pressure in about four minutes.

β) Effect on the Heart

(a) Effect on the Electrocardiogram

In dogs, at an intravenous dose of 5 mg/kg, rubidomycin did not cause any change in the electrocardiogram or in the heart rate. The period of observation extended over five hours after the administration of the product.

In guinea-pigs urethanized at an intraperitoneal dose of 1 g/kg and weighing about 300 g, the continuous intravenous perfusion of rubidomycin in the form of a 0.5% solution and at a rate of 1 ml/min may cause, a few minutes before death (which generally occurs 15 to 20 minutes after perfusion is begun), various anomalies in the electrocardiogram: bradycardia, atrioventricular dissociation and ventricular extrasystoles preceding cardiac arrest.

(b) Effect on the Myocardiogram

In dogs, at an intravenous dose of 5 mg/kg, rubidomycin had no effect on cardiac contractility [155]. The period of observation extended over five hours after the administration of the product.

(c) Effect on the Papillary Muscle

Rubidomycin had no depressor effect, at cumulative concentrations in the range 0.01 to 100 mg/liter, on the papillary muscle of the heart of the cat.

b) Effect on Respiration

In dogs, at an intravenous dose of 5 mg/kg, rubidomycin had practically no effect on the respiratory rate or output. The period of observation extended over three hours after the administration of the product. The only effect noted was a temporary increase of 40% for three minutes, on injection, in the amplitude of the respiratory movements, followed by a slight decrease of 25% and a return to normal in about 20 minutes.

c) Effect on the Sympathetic Nervous System

In dogs, at an intravenous dose of 5 mg/kg, rubidomycin showed no marked effect on the sympathetic nervous system (hypertension was induced by the intravenous injection of adrenalin and noradrenalin and by the bilateral occlusion of the carotid arteries) and the parasympathetic nervous system (hypotension was induced by the intravenous injection of acetylcholine and by stimulation of the peripheral vagal ending). The period of observation extended over five hours after the administration of the product.

It may be concluded from these tests that rubidomycin is of greatest value as an antitumor agent [80, 164, 165, 224, 225] in animals, being effective against solid tumors (sarcomas and carcinomas), Ehrlich ascites tumor, and the leukoses.

Chapter 4

Fixation in the Cell and Mechanism of Action

1. Fixation in the Cell

Except for recent papers by DI MARCO et al., one of which [67] deals with the *in vitro* investigation of its metabolization by tissue extracts, the other [65] with blood and urine determinations in the animals treated, no work has been published, as far as we know, on the analysis of rubidomycin in animal tissues.

We have attempted to analyze rubidomycin in biological fluids and tissues of mice, rats, and rabbit. The results obtained showed that the substance isolated and analyzed is actually a mixture of the antibiotic itself and of one its metabolites. The separation and analysis of these two substances are at present in progress.

We have, however, studied the cell fixation of rubidomycin and of its metabolites as a whole, using on the one hand tritiated rubidomycin and carrying out both an autoradiography and measuring the total radioactivity of the chief cell constituents and, on the other hand, applying the technique of fluorescence.

a) Autoradiography of KB Cells

We incorporated tritiated rubidomycin in KB cells. Tritiated rubidomycin (in the titer of $1 \mu g = 10^5$ d.p.m. [1]) was added up to a concentration of $5 \mu g/ml$ to a slide culture of cells. After a period of 60 minutes (if the period is shorter the labeling is less satisfactory), the slides were removed, fixed with methanol, and treated for 10 days with Ilford K_2 emulsion. They were then developed and fixed, and the cells were stained with May-Grünwald-Giemsa stain and examined microscopically with an immersion objective (Fig. 20). As a control, we used concurrently a tritiated substance which selectively fixes itself on the nucleus: tritiated thymidine (Centre d'Etude de l'Energie Nucléaire, Mol, Belgium), at a concentration of $1 \mu Ci/ml$. Fig. 21 illustrates the expected nuclear fixation of labeled thymidine. As for the tritiated rubidomycin, it should be pointed out that it too is fixed mainly in the nucleus and to a much lesser extent in the cytoplasm.

b) Measurement of the Tritiated Rubidomycin in Normal Hepatic Cells

Radioactivity measurements were carried out in normal hepatic cells of rats treated intravenously with 1 mg/kg of tritiated rubidomycin and sacrificed 2 and 24 hours later.

Table 16. *Radioactivity of rubidomycin fixed in normal hepatic cell constituents*

Cell constituents	Radioactivity (d.p.m./g of fresh tissue)	
	Rats sacrificed after 2 h	Rats sacrificed after 24 h
Nuclei	$2.50 \cdot 10^{-5}$	$2.34 \cdot 10^{-4}$
Mitochondria	$0.90 \cdot 10^{-5}$	$1.17 \cdot 10^{-4}$
Microsomes	$0.02 \cdot 10^{-5}$	$0.03 \cdot 10^{-4}$

---

[1] d.p.m. = disintegrations per minute.

After separation of the main cell constituents by the standard method of fraction centrifugation, the radioactivity of the various fractions was measured.

The results, which are shown in Table 16, show that, as in the case of KB cancer cells, the greater part of the antibiotic or of its metabolites is fixed in the nuclear fraction of the normal cell.

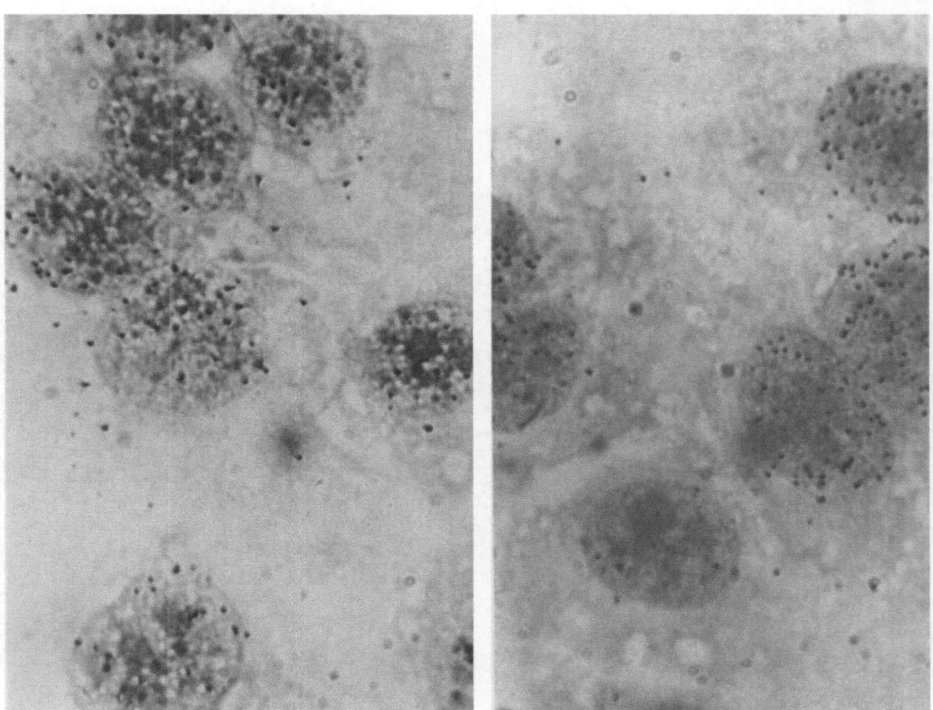

Fig. 20. Autoradiograph of KB cells in the presence of tritiated rubidomycin.
May-Grünwald-Giemsa stain; magnification ×1100

Fig. 21. Autoradiograph of KB cells in the presence of tritiated thymidine.
May-Grünwald-Giemsa stain; magnification ×1100

c) Fluorescent Microscopy (KB Cells)

KB cells grown on slides and left in contact for 10 minutes with a very high concentration (50 µg/ml) of rubidomycin were examined under a fluorescent microscope.

Fluorescence was well-marked in the cytoplasm and nucleolus, scanty or non-existent in the nucleus (Fig. 22 and 23). Our interpretation of this paradoxical result is as follows: the rubidomycin becomes fixed on the DNA of the nucleus and its fluorescence greatly diminishes or disappears. For this reason the nucleus is not fluorescent, whereas the cytoplasm and the nucleolus, which contain free rubidomycin, are clearly defined. If the rubidomycin concentration is reduced (5 µg/ml) no fluorescence is observed in the cell; the antibiotic has penetrated into the cell but has become selectively fixed to the DNA of the nucleus, and as a result of this union has lost its fluorescent property.

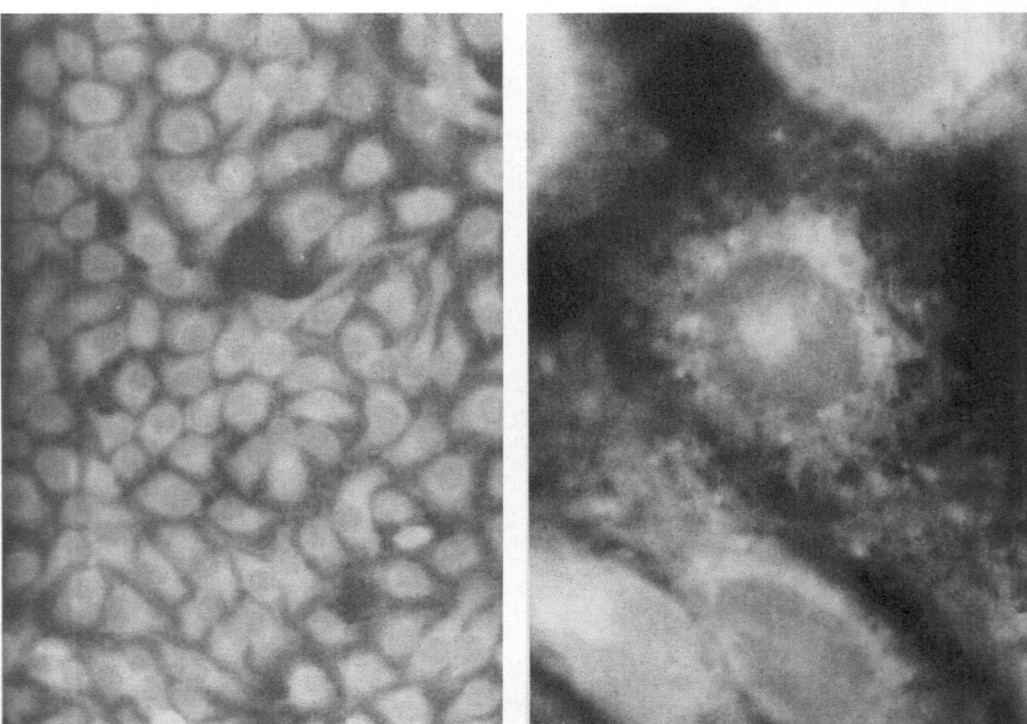

Fig. 22. Fluorescence of KB cells in the presence of rubidomycin. Magnification ×250
Fig. 23. Fluorescence of KB cells in the presence of rubidomycin. Magnification ×1100

2. Mechanism of Action

a) Action on Energy Metabolism

Like ionizing radiation, many cytostatic substances belonging to very different chemical families (mustards, ethyleneimine derivatives, some antibiotics, etc.) depress cell respiration and, in weak concentration, inhibit the processes of glycolysis, which are the main generators of the energy needed for the growth and reproduction of cancer cells.

The action of most of these substances on energy mechanisms seems to be linked to their effect on the intracellular level of coenzyme I (nicotinamide adenine dinucleotide = NAD); when this is reduced below a certain threshold, glycolysis stops. This particular action would seem to be, *inter alia*, one of the main factors of the anticancer activity of these substances [232].

We studied the *in vitro* action of rubidomycin on energy metabolism by measuring its effect on respiration, glycolysis and, occasionally, the NAD level of the tumor cell in Ehrlich ascites.

α) Method of Study

Ascitic fluid was taken from mice 7—10 days after transplantation of Ehrlich tumor. After centrifugation at 450 g[2] for five minutes, the cells were cleared of red blood cells by Chance and Hesse's method [54] and washed with physiological saline.

The cells were then again centrifuged and placed in suspension, either in Krebs-Ringer phosphated buffer solution at a pH of 7.4 to measure respiration or else in Krebs-Ringer bicarbonated buffer solution at a pH of 7.0 to measure aerobic and anaerobic glycolysis. The measurements were carried out by conventional methods [222] with the Warburg apparatus. At the beginning of the test, rubidomycin dissolved in these solutions was added to the cell suspension, which contained 8 to 16 mg of cells per ml (dry weight) according to the case. Glucose was added, either at the same time as the rubidomycin or 40 minutes afterwards. The measurements were carried out every 10 minutes for one hour, 40—60 minutes after the addition of the product, after gassing, and once thermal equilibrium of the suspensions was achieved.

To measure coenzyme I, the cell suspensions were centrifuged at 2500 g for 15 minutes at 0° C. The cells were homogenized in 10 ml 0.33 N perchloric acid and, after being left for 30 minutes in the cold, the homogenates thus obtained were centrifuged as before. The supernatant fluids were cleared of their content of rubidomycin by several successive extractions with butanol. They were then directly used for the measurement of NAD by Lowry et al.'s method [156] or Weitzel and Buddecke's method [231].

The respiration and glycolysis intensities were expressed by the usual coefficients $(Q_{O_2}, Q_{CO_2}^{O_2}, Q_{CO_2}^{N_2})$, and the NAD levels in nanomoles per mg of cells (dry weight).

β) Results

From the results shown in Tables 17 and 19, it can be seen that rubidomycin does not appreciably affect either respiration or glycolysis in cancer cells, even at the relatively very high concentration of 120 µg/ml. This concentration is much higher than that which inhibits the growth of tumors in animals and can never be attained, even in courses of treatment of very long duration.

As may be seen in Table 21 (p. 61), rubidomycin lowers the level of NAD in proportion to its concentration. There is no question of this being an artefact appearing during the measurements and perhaps caused by the antibiotic's fluorescence interfering with that of the NAD, since the fall in the level of the coenzyme is much less if nicotinamide is added to the cell suspension (Table 18).

In the same experimental conditions, triaziquone, known for its intense effect on glycolysis in cancer cells, reduced the NAD content to zero and completely inhibited aerobic and anaerobic glycolysis at markedly weaker concentrations—of the order of some µg/ml.

---

[2] Jouan G 60 centrifuge, at 2000 revolutions per minute.

Table 17. *Action of rubidomycin on respiration and glycolysis in Ehrlich ascitic cells*

Concentration of drug in µg/ml		Respiration (a)		Glycolysis (b)	
		endogenous Q_{O_2}	with glucose Q_{O_2}	aerobic O_2 Q_{CO_2}	anaerobic N_2 Q_{CO_2}
Controls		6.50	3.00	20.65	44.00
Rubidomycin	240	4.95	3.90	12.30	32.10
	120	6.10	3.70	20.30	38.00
	60	6.00	3.25	20.50	39.90
	30	5.90	2.75	19.00	40.60

(a) glucose added 40 minutes after rubidomycin
(b) glucose added at same time as rubidomycin

Respiration:

Main space:	2.4 ml cell suspension containing 0.05 ml cells
Central space:	0.2 ml KOH at 20%
Diverticulum:	0.4 ml Krebs-Ringer phosphate buffer (endogenous respiration) or 0.4 ml 0.1 M glucose in phosphate buffer
Gaseous phase:	O_2: 93%, CO_2: 7%

Glycolysis:

Main space:	2.4 ml cell suspension containing 0.05 ml cells
Central space:	0.2 ml Krebs-Ringer bicarbonate solution
Diverticulum:	0.4 ml 0.1 M glucose in bicarbonate buffer
Gaseous phase:	O_2: 93%, CO_2: 7% (aerobic glycolysis) N_2: 95%, CO_2: 5% (anaerobic glycolysis)

Table 18. *Action of nicotinamide on the fall in cell NAD caused by rubidomycin (aerobic)*

Rubidomycin µg/ml	0 (controls)		30		60		120	
Nicotinamide mg/ml	0	1.8	0	1.8	0	1.8	0	1.8
NAD levels: nanomoles/mg cells (dry weight)	3.7	4.3	2.3	3.6	1.8	3.7	1.3	3.1
As percentage of controls			62.1	83.7	48.6	86.0	35.1	72.0

At *in vitro* concentrations actively affecting tumor development, rubidomycin thus has no specific effect on energy-generating systems. This same finding was made by DI MARCO et al. [70]. Since the cytostatic action of the compound cannot be accounted for by its feeble effect on the energy metabolism of the cell, it must be sought elsewhere.

b) Action on Protein Synthesis

The action of rubidomycin on protein synthesis in the cancer cell has been studied *in vitro* by measuring the rate at which ^{14}C-alanine was incorporated into the proteins.

α) *Method of Study*

Ascitic cell suspensions in Krebs-Ringer bicarbonated buffer solution with glucose and containing known amounts of rubidomycin were prepared by the method described on page 49.

Forty minutes after the rubidomycin was added, to 3 ml of the suspensions was added 0.1 ml of a solution containing 25 µCi of ^{14}C-alanine dissolved in the buffer. The ^{14}C-alanine was obtained from the Commissariat à l'Energie Atomique, Saclay, and had the specific activity of 3.7 mCi/millimole. After incubation for 30 minutes at 37° C, during which aerobic and anaerobic glycolysis were measured, the containers were placed on ice to stop reactions. The cells were separated off by centrifugation, then homogenized in 0.33 N perchloric acid. The homogenates were left for 30 minutes at 0° C, then centrifuged, and the protein residue obtained was washed twice in 10 ml ethanol and once in 10 ml alcoholic ether. The solvent was removed by centrifuging, the precipitate was dissolved in 2 ml N sodium hydroxide, and the volume of the solution obtained was brought up to 20 ml with distilled water. The solution was left for 12 hours at laboratory temperature, then two aliquots of 0.2 ml were taken, one to measure the proteins by Lowry et al.'s method [157], the other to measure the radioactivity. The radioactivity, as expressed in d.p.m., was measured with the 33.14 Packard-Tricarb spectrometer, after neutralization of the alkaline solution by an ethanolic solution of 0.01 N hydrochloric acid and addition to the volume of 20 ml of toluene containing per liter 4 g 2.5-diphenyl-oxazole and 0.1 g 1.4- bis (4-methyl-5-phenyl-2-oxazolyl) benzene (scintillation fluid).

β) *Results*

Table 19 shows that rubidomycin does not inhibit the incorporation of labeled alanine in proteins at concentrations equal to or less than 60 µg/ml. It should be noted that these concentrations, which have no influence whatever on glycolysis, reduce the cell NAD content in the same experimental conditions by 30% to 50%.

Table 19. *Action of rubidomycin on incorporation of ^{14}C-alanine in Ehrlich ascitic cell proteins* (30 minutes incubation with labeled precursor)

Drug concentration µg/ml		Aerobic (average of three tests)		Anaerobic (average of three tests)	
		$Q_{CO_2}^{O_2}$	^{14}C protein radioactivity d.p.m. 10^{-3}/mg protein	$Q_{CO_2}^{N_2}$	^{14}C protein radioactivity d.p.m. 10^{-3}/mg protein
Controls		22.7	6.9	44.4	6.0
Rubidomycin	120	21.8	4.6	38.1	5.5
	60	22.7	7.4	39.3	5.3
	30	22.0	7.2	40.6	5.2
	15	21.8	7.0	41.8	6.5

Honig et al. [125] have also shown that, in the presence of glucose, rubidomycin does not, at a concentration of 50 µg/ml, inhibit protein synthesis in sarcoma 37 cells (ascitic form).

However, at twice the concentration (120 µg/ml), but only aerobically, a fall in the incorporation rate of about 30⁰/₀ was observed, though there was no change in the intensity of glycolysis, implying a continuously adequate production of ATP.

Given the high concentrations needed to produce an appreciable fall in the incorporation of ^{14}C-alanine, the antimitotic effect of rubidomycin cannot be attributed to its action on protein synthesis.

c) Action on the Synthesis of Nucleic Acids

The theory might be advanced that rubidomycin acts on the nucleic acids, because it induces lysogenesis in *Escherichia coli* [117, 181] and has an antiphage action [117, 177, 191].

The action of rubidomycin on nuclear synthesis in the cancer cell was studied by three methods. The first was by observing DNA and RNA changes by selective staining with the fluorescent microscope, the other two were by assessing the incorporation of thymidine and tritiated uridine in the nucleic acids, on the one hand by autoradiography, on the other by measurement of the radioactivity of the tritium incorporated in the cell RNA and DNA.

α) Staining of the Cells with Acridine Orange

Mice with Ehrlich ascitic tumor were injected intraperitoneally once, on the ninth day after the graft, with the compound (1.25 mg/kg). After treatment ascitic fluid was taken at intervals of 1, 4, 7, and 24 hours, smeared on a slide, stained with acridine orange [31, 32, 221], and examined by fluorescent microscope. In all 2000 cells were examined per slide and they may be classified as follows:

  I: normal cells
 II: cells with RNA changes (green nucleus, pale red cytoplasm)
 II a: cells with DNA changes (pale green nucleus, orange-red cytoplasm)
 III: cells with RNA and DNA changes (pale green nucleus, pale orange cytoplasm)
 IV: cells with very marked RNA and DNA changes (nucleus little stained, cytoplasm unstained).

With the percentage determined for each cell category a histogram (Fig. 24) was constructed. It can be seen from the histogram that rubidomycin has a harmful action, of rapid onset and prolonged duration, on the nucleic acids (RNA and DNA). In this experimental system actinomycin D has an action akin to that of rubidomycin.

β) Autoradiographic Study of the Incorporation of Tritiated Thymidine and Uridine into the Treated Cells

(a) Method of Study

Rubidomycin concentrations of 0.2 and 1.0 µg/ml were chosen for the autoradiographic investigations. After 2, 6, and 24 hours' reaction with rubidomycin, cultures of HeLa cells were transferred to plastic containers for the radioactive test phase. 1.0 µCi/ml ^{3}H-thymidine (spec. activity 20 600 mCi/mMol) and 1.0 µCi/ml ^{3}H-uridine (spec. activity 31 000 mCi/mMol) were added to the nutrient solution containing the appropriate rubidomycin concentration. The radioactive test phase lasted for one hour, so that the cultures were exposed to the action of rubidomycin for a total of 3, 7, and 25 hours. Parallel to this, the untreated control cultures in rubidomycin-free medium were labeled. For fixation purposes 10% formalin was used for one hour at

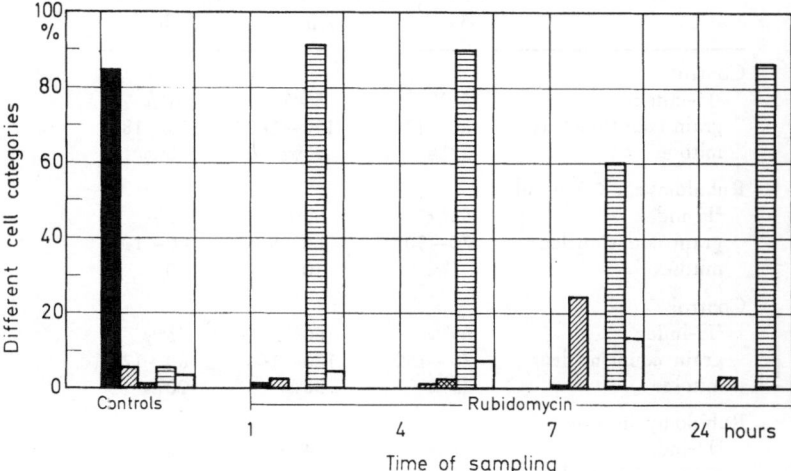

Fig. 24. Histogram showing staining with acridine orange of Ehrlich ascites cells after intraperitoneal administration to mice.
Categories of cells ■ ▨ ▨ ▤ ▢
I II IIa III IV

room temperature with the addition of 0.5 g/liter of the inactive precursor and 80% alcohol. The preparations were then placed in running water for one hour. The stripping film technique (Kodak AR 10) was used for the autoradiographic procedure.

The exposure time was seven days. The parameters examined were the mitotic index, the ³H-thymidine labeling index (n = 2000), and the silver grain count per nucleus (n = 100, on account of the high silver grain density only orientating). In the tests with ³H-uridine, in addition to the silver grain count, the nuclear area was determined with an ocular micrometer in relative units in order to allow for the variable size of the nucleus.

(b) Results

(α) ³H-thymidine

The effect of rubidomycin on DNA metabolism and on the mitotic index is illustrated in Table 20. At a concentration of only 0.2 µg/ml, a reduction in mitotic index from 22 to 1‰ occurred after only three hours. During the remainder of the experiment no further mitoses were observed. After seven hours, and even more after 25 hours, an increase in the ³H-thymidine labeling index was found for the treated cultures, while the silver grain count per labeled nucleus was markedly reduced compared to the controls, as can be seen in the autoradiograms in Fig. 25. After seven hours at a concentration of 1.0 µg/ml (Table 20) a reduction in the labeling index was found, the labeled cells showing very small silver grain counts per nucleus, as had already been found after 3 hours (Fig. 26). After 25 hours the incorporation of ³H-thymidine was no longer demonstrable (Table 20).

Table 20. *HeLa cells after treatment with rubidomycin and labeling with ³H-thymidine* (1 μCi/ml)

Product	Duration of treatment		
	3 h	7 h	25 h
Control			
³H-index	50%	50%	51%
grain count/nucleus	80—130	120—160	150—180
mitoses	22‰	23‰	30‰
Rubidomycin 0.2 μg/ml			
³H-index	53%	59%	75%
grain count/nucleus	70—90	60—90	60—100
mitoses	1‰	0	0
Control			
³H-index	63%	60%	43%
grain count/nucleus	70—150	120—160	80—120
mitoses	25‰	30‰	20‰
Rubidomycin 1 μg/ml			
³H-index	51%	24%	0
grain count/nucleus	6—20	6—20	0
mitoses	1‰	0	0

(β) ³H-uridine

At a concentration of 0.2 μg/ml, at first only slight deviations of the average silver grain count per nucleus were found. After 25 hours (Fig. 27) the nuclei were much larger and the silver grain count per nucleus was almost equal to that of the controls. The silver grain density had therefore diminished. Fig. 28 shows the corresponding autoradiogram. At a concentration of 1 μg/ml a marked reduction in silver grain count was found after three hours, and it was even more marked after seven hours (Fig. 29 and 30). After 25 hours the cells that still remained alive showed marked variations in size and silver grain count.

Rubidomycin caused inhibition of mitotic activity at a concentration of 0.2 μg/ml— a concentration in which, even after 24 hours of exposure to the substance *in vitro*, still no notable lethal damage was found. After only three hours the mitotic index had fallen from 22 to 1‰. The fact that the ³H-thymidine index did not fall during this short time but rather increased (Table 20) indicates that a direct attack on the G_2-phase, i. e., the phase between DNA synthesis (S-phase) and mitosis, was taking place. At this same concentration of 0.2 μg/ml of rubidomycin, the silver grain count after ³H-thymidine labeling was significantly reduced per labeled nucleus (Table 20 and Fig. 25). Since the time of exposure in both experiments was the same, this finding indicates a reduction in the rate of DNA synthesis in the individual nuclei. Therefore, the S-phase was impaired by rubidomycin at the same concentration as the G_2-phase. As the exposure time to rubidomycin increased, so did the percentage of ³H-thymidine labeled cells from 53% after three hours to 75% after 25 hours. This finding suggests that the total duration of DNA synthesis is prolonged at the same time as the rate of DNA synthesis is reduced, but this is valid solely on the assumption that only a few cells died within the period of observation. At this concentration of 0.2 µg/ml the lethal damage (Fig. 11) is only 7.5⁰/o after 24 hours and can therefore be disregarded.

Since the initiation of mitotic cell division was completely inhibited after no more than seven hours of exposure to rubidomycin (Table 20), the cells appear to persist in the S-phase or G_2-phase. The mechanism of action of rubidomycin is, therefore, in the main not different from that of the other cytostatics [182, 232].

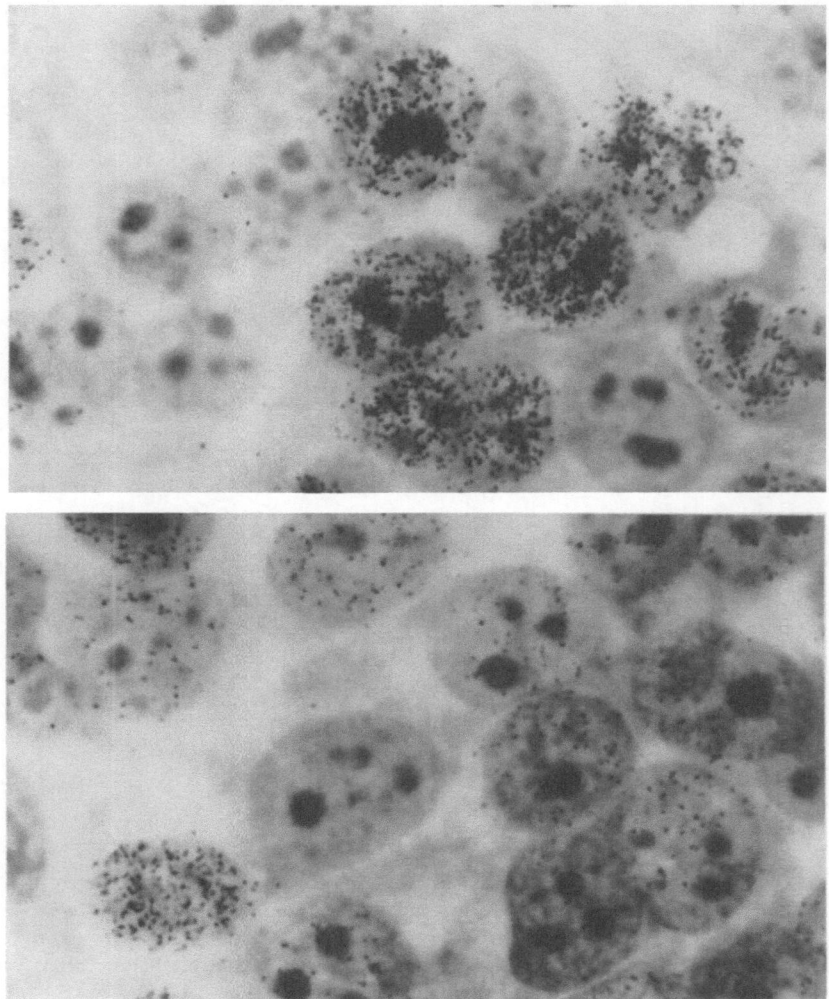

Fig. 25. HeLa cell culture labeled with ³H-thymidine. Above: control; Below: after 25 hours exposure to 0.2 µg/ml rubidomycin (magnification ×1200)

At the low concentration of 0.2 µg/ml of rubidomycin, the only cytomorphological change observable, in spite of the above-mentioned serious interference with DNA metabolism, consisted of no more than a slight raggedness of the nucleoles. The amount of ³H-uridine taken up per nucleus at this dosage was still quantitatively unchanged after 25 hours, but was markedly reduced when referred to the nuclear area since the nuclei had become enlarged (Fig. 27 and 28). Only after seven hours of exposure to 1 µg/ml (Fig. 13 b) were morphological changes apparent: the karyo-

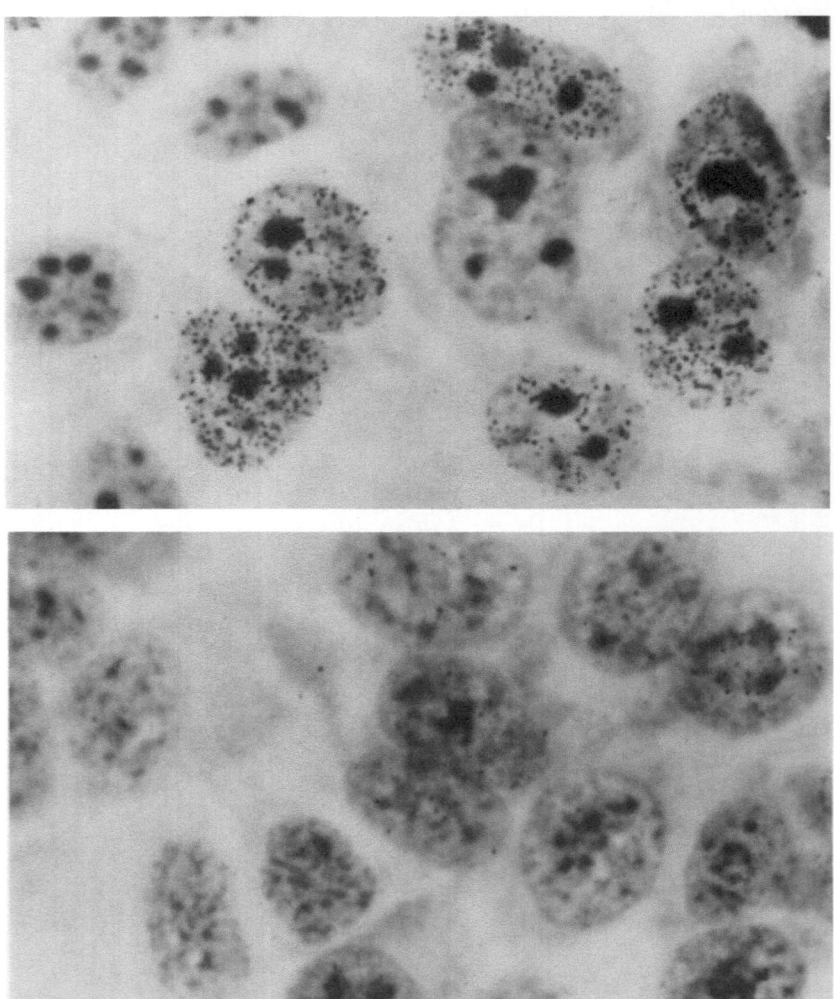

Fig. 26. HeLa cell culture labeled with ³H-thymidine. Above: control; Below: after 3 hours exposure to 1 µg/ml rubidomycin (magnification ×1200)

plasm was partly loosened and partly lumpy, and the nucleoles had shrunk. At the same concentration, and after the same length of time, the silver grain count after the incorporation of ³H-uridine was significantly reduced (Fig. 29 and 30). Nevertheless, ³H-uridine was still taken up in measurable quantities.

If, as far as it is possible to draw conclusions from *in vitro* studies, the action of rubidomycin on the DNA and on the RNA metabolism of HeLa cells is compared,

DNA synthesis is found to be inhibited at lower concentrations than RNA synthesis, and interference with DNA metabolism is more pronounced.

Rubidomycin inhibits the synthesis of the nucleic acids. Its action is very marked on DNA synthesis but less so on RNA synthesis, as DI MARCO et al. [74] have also

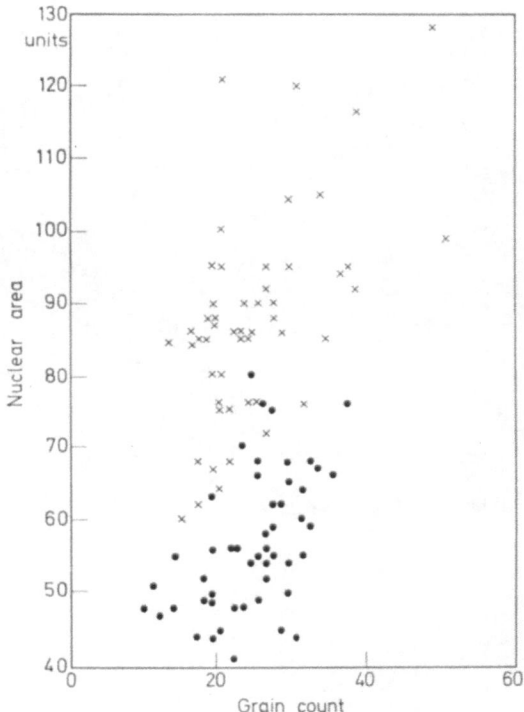

Fig. 27. Nuclear area and silver grain count of HeLa cells after labeling with ³H-uridine (0.2 μg/ml rubidomycin, 25 hours). • Controls; × Cells treated

shown using HeLa cells, and PARISI and SOLLER [177] and SANFILIPPO and MAZZOLENI [191] with DNA and RNA phages. We demonstrated that, in KB cell cultures and at concentrations that were not cytotoxic, rubidomycin inhibits the development of vaccinia virus (a DNA virus) and has no effect on vesicular stomatitis virus (an RNA virus). The observations of BLUMBERG and KALAMOVA [34] fully confirm our results, for they found that the effect on vaccinia virus was favorable and that there was no effect on myxovirus (type A₂ influenza virus) and some other RNA viruses.

γ) Determination by Measurement of the Incorporation of Tritiated Thymidine and Uridine in the Nucleic Acids of Ehrlich Ascitic Tumor Cells

The action of rubidomycin on the incorporation of tritiated thymidine and uridine into nucleic acids was determined *in vitro* by standard physical and chemical methods. To relate such action to that of rubidomycin on energy metabolism, the intensity of glycolysis and the NAD level in the cells were measured simultaneously.

(a) Method of Study

The method employed is described in the previous paragraph. The tests, however, were only performed aerobically and the labeled alanine was replaced by tritiated thymidine and uridine, with 1 μCi of radioactive substance per 3 ml of cell suspension.

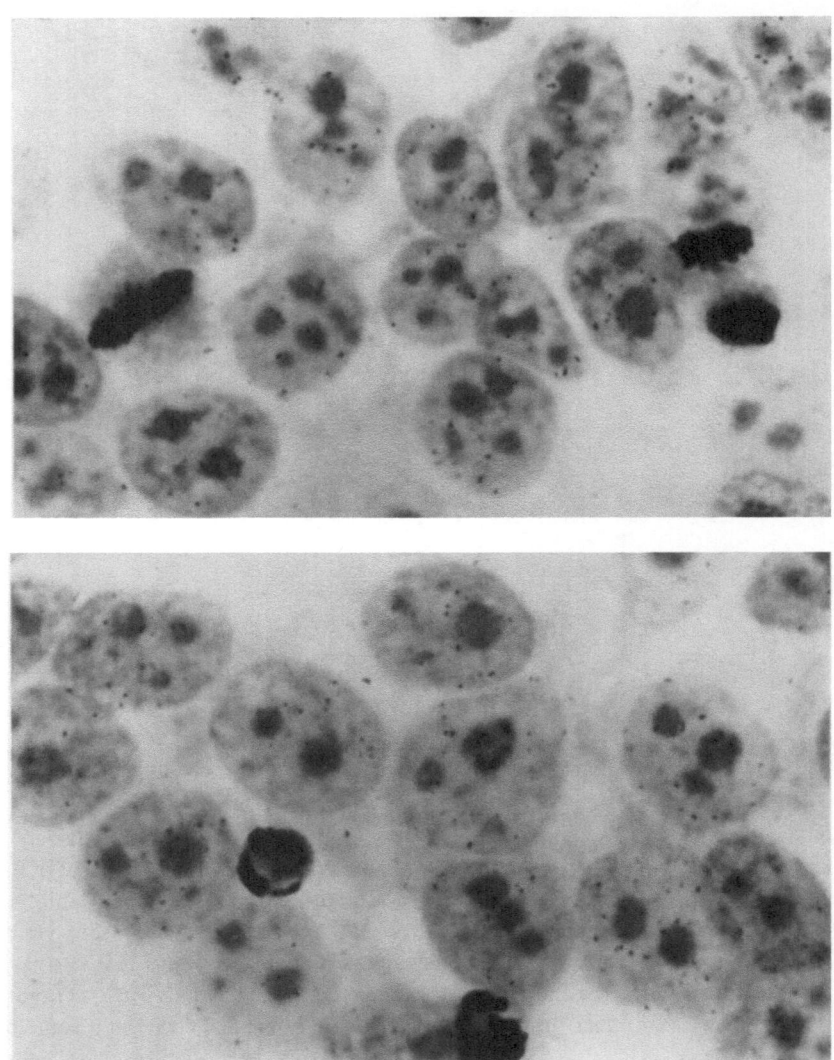

Fig. 28. HeLa cell culture labeled with ³H-uridine. Above: control; Below: after 25 hours exposure to 0.2 μg/ml rubidomycin (magnification ×1200)

The ³H-thymidine came from the Radiochemical Centre (Amersham) and its specific activity was 1.44 Ci/mMol; the ³H-uridine came from the Commissariat à l'Energie Atomique (Saclay) and its specific activity was 3.3 Ci/mMol. The cell suspensions were incubated for 30 minutes with the labeled precursors and treated exactly as described previously up to the stage of obtaining proteins free from lipids. The nucleic acids were extracted from the proteins by SCHMIDT and THANNHAUSER's method [193] and their amounts determined by ultraviolet spectrophotometry and phosphorus

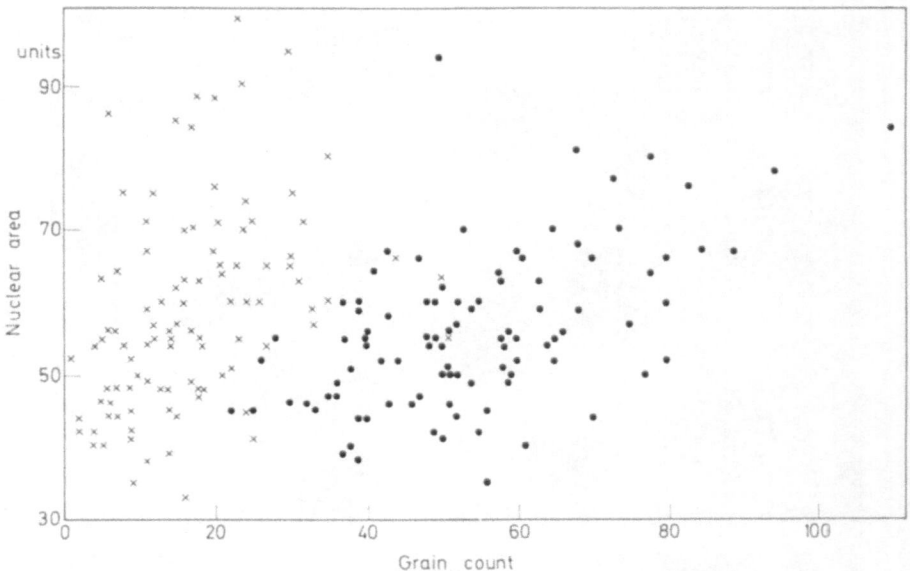

Fig. 29. Nuclear area and silver grain count of HeLa cells after labeling with ^{3}H-uridine (1 µg/ml rubidomycin, 7 hours). ● Controls; × Cells treated

measurement by KING's method [146]. The NAD was measured by the method mentioned above [156] in the supernatant fluid obtained after centrifuging the perchloric cell homogenate

The radioactivity of the perchloric DNA and RNA solutions, expressed in d.p.m., was measured as in the case of the labeled proteins, but after preliminary neutralization of the aliquots (0.3 ml) with N/40 alcoholic sodium hydroxide and addition of scintillation fluid to the appropriate volume (about 20 ml).

(b) Results

The results of these short-term trials (Table 21), which with few differences confirmed those of autoradiography, show that, at a concentration of 15 µg/ml rubidomycin inhibits equally and to an important extent (more than 50%) the incorporation of tritiated thymidine and uridine into the nucleic acid of cancer cells.

At a concentration four times greater (60 µg/ml), the inhibition became nearly 85%. Glycolysis, however, was unaffected, although the NAD level fell by only 30% and protein synthesis, as measured during other tests carried out in the same experimental conditions, remained unimpaired.

Only at the relatively high concentration of 120 µg/ml did rubidomycin cause an appreciable fall in the NAD level (55%), while incorporation of thymidine into the DNA was almost completely inhibited. In these conditions, whatever the process put forward to explain the fall (blocking of synthesis or activation of the degradation of the coenzyme), there is no parallelism such as HILZ et al. [122] observed in other cytostatic drugs between the fall in NAD and the inhibition of DNA synthesis.

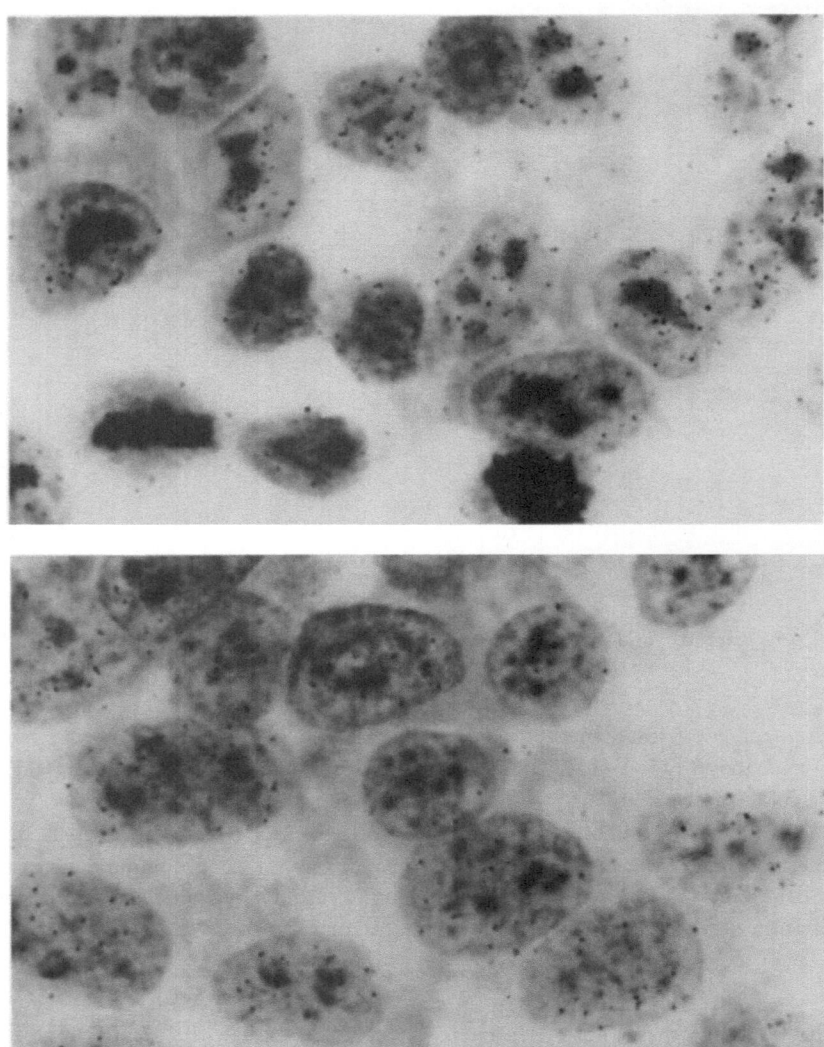

Fig. 30. HeLa cell culture labeled with ³H-uridine. Above: control; Below: after 7 hours exposure to 1 µg/ml rubidomycin (magnification ×1200)

Fig. 31 indicates that rubidomycin has a stronger action on nucleic acid synthesis than on protein synthesis in the Ehrlich ascitic cell. In these circumstances, it seems that this, its main activity, is the only one responsible for its cytostatic capacity.

Early and marked inhibition by rubidomycin of the incorporation of tritiated precursors into DNA and RNA has been observed *in vitro* and *in vivo* in certain micro-organisms [11, 118, 228] and in animal cells [72, 81, 125, 189, 190, 192, 201,

Table 21. *Action on aerobic glycolysis, NAD level, and specific radioactivity of DNA and RNA in Ehrlich ascitic tumor cells*

(^{3}H-thymidine or ^{3}H-uridine added 40 minutes after rubidomycin; incubation period 30 minutes)

Rubidomycin concentration µg/ml	(0) (controls)	15	30	60	120
Aerobic glycolysis $Q_{CO_2}^{O_2}$ (2 tests)	15.50	15.70	14.85	14.50	15.30
NAD nanomoles/mg cells (dry weight) (2 tests)	2.13	2.00	1.81	1.61	0.95
Specific activity of DNA (^{3}H-thymidine) d.p.m./µg P-DNA (1 test)	666	345	232	64	14.5
Specific activity of RNA (^{3}H-uridine) d.p.m./µg P-RNA (1 test)	106	56	25.6	15.5	8.7

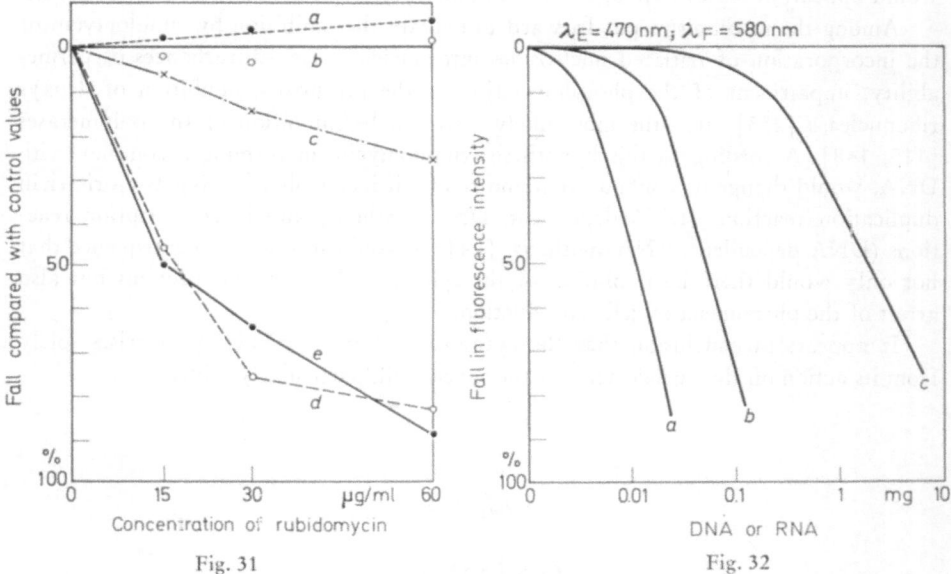

Fig. 31   Fig. 32

Fig. 31. Effect on aerobic glycolysis and on the synthesis of proteins and nucleic acids. a Incorporation of ^{14}C-alanine in proteins; b $Q_{CO_2}^{O_2}$ (aerobic glycolysis); c NAD level; d Incorporation of ^{3}H-thymidine in DNA; e Incorporation of ^{3}H-uridine in RNA

Fig. 32. Effect of DNA and RNA on the fluorescence of rubidomycin. Solutions of rubidomycin: a Solution (5 µg) in 0.15 M sodium chloride plus DNA; b Solution (15 µg) in buffer at pH 7.0 plus DNA; c Solution (15 µg) in buffer at pH 7.0 plus RNA

202]. It was recently shown that it also inhibits, *in vitro* and *in vivo*, the incorporation of ^{32}P into the nucleic acids of normal and of cancerous tissues [218]. This implies its direct or indirect action on the enzyme systems responsible for nucleic acid synthesis, though it does not explain the mechanism or indicate the sites of action.

Red staining of the protein precipitate obtained during the isolation of the nucleic acids shows that rubidomycin penetrates easily and is very largely fixed in the cells. It is therefore probable that, like actinomycins and other antibiotics, rubidomycin becomes fixed on the nucleoproteins and particularly on the nucleic acids, to form complexes with them that, when present in synthesizing systems, may render the latter non-functioning.

The formation of such complexes has in fact been shown by the study of changes in the physicochemical properties of nucleic acids in the presence of rubidomycin [51, 100, 144, 145]. The hypothesis has been expressed that it may become fixed on purine and pyrimidine bases [188] and insert itself into the DNA helix [104, 228] to produce stable complexes. We, however, have merely studied the variations in the fluorescence of rubidomycin (5 and 15 µg), dissolved in 0.15 M sodium chloride or in Krebs-Ringer bicarbonated buffer solution in the presence of increasing amounts of yeast RNA or DNA extracted from ascitic cells by KAY et al.'s method [143].

The results of that study (Fig. 32) showed that the fluorescence of rubidomycin solutions, as measured by the Aminco-Bowman spectrofluorimeter, was relatively little changed in the presence of RNA, but was on the other hand greatly reduced in the presence of low concentrations of DNA. In these circumstances, rubidomycin would appear, at least *in vitro*, to have an almost selective action in relation to DNA.

Among the mechanisms put forward to explain the inhibition by rubidomycin of the incorporation of tritiated nucleosides into nucleic acids—disturbances of permeability, impairment of the phosphorylation of the precursors, inhibition of deoxyribonuclease [233] etc.—the most likely seems to be inhibition of the polymerases [115, 148]. According to this hypothesis, rubidomycin, in forming a complex with DNA, would change its configuration and render it incapable of acting as a primer in duplication reactions (DNA-dependent, DNA-synthesis) and in transcription reactions (DNA-dependent, RNA-synthesis) [64]. It would follow as a consequence that not only would there be inhibition of the synthesis of RNA and proteins but also arrest of the phenomena of cell reproduction.

It appears in conclusion that the cytostatic action of rubidomycin arises solely from its action on the nucleic acids of the cancer cell, particularly DNA.

Chapter 5

Toxicology

The acute toxicity of rubidomycin was determined in several animal species (mouse, rat, guinea-pig, rabbit, and dog) and its subacute toxicity in the mouse.

Its long-term (three months) toxicity was also studied in the rabbit and dog.

Finally, we looked for the possible existence of a teratogenic effect on the chick embryo, mouse, and rabbit, and of a carcinogenic effect on the mouse.

Table 22. *Acute and subacute toxicity*

Test	Species Strain Sex	Body weight	Unit vol. admin.[a]	i.v. Dose mg/kg	i.v. Mortality[b]	i.v. LD$_{50}$	i.p. Dose mg/kg	i.p. Mortality[b]	i.p. DL$_{50}$	p.o. Dose mg/kg	p.o. Mortality[b]	p.o. DL$_{50}$
Acute toxicity	Mice CD-1 (COBS)[c] males and females	16—20 g	50 ml/kg	40	10 d/10	app. 20	6.7	10 d/10	app. 5	450	8 d/8	app. 300
				27	7 d/10		4.5	3 d/10		300	4 d/8	
				18	1 d/10		3	0 d/10		200	0 d/8	
				12	2 d/10		2	0 d/10		130	0 d/8	
				8	2 d/10							
	Rats CD (COBS)[c] males and females	115—125 g	5 ml/kg	25	5 d/5	12.5—25				300	2 d/5	of order of 300
				12.5	0 d/5					150	0 d/5	
				6.2	0 d/5					75	0 d/5	
	Guinea-pigs (common species) males and females	220—260 g	5 ml/kg	25	5 d/5	app. 6						
				12.5	5 d/5							
				6.2	3 d/5							
				3.1	0 d/5							
	Rabbits (common species) males and females	1.8—2.7 kg	1 ml/kg	25	3 d/3	app. 5						
				12.5	3 d/3							
				6.2	2 d/3							
				3.1	0 d/3							
	Dogs (common species) males and females	5.5—11.2 kg	0.5 ml/kg	8	2 d/2	app. 4				40	3 d/3	20—40
				4	1 d/2					20	0 d/2	
				2	0 d/2					10	0 d/2	
				1	0 d/2							
Subacute toxicity (5 days of daily treatment)	Mice CD-1 (COBS)[c] males and females	16—20 g	50 ml/kg	10	6 d/10	app. 10	3.2	8 d/8	app. 2			
				6.7	0 d/10		2.5	7 d/8				
				4.5	0 d/10		2	4 d/8				
				3	1 d/10		1.6	0 d/8				
				2	0 d/10		1.2	0 d/8				

[a] At 1 ml/min when administered intravenously.
[b] No. of animals dead in relation to no. utilized 15 days after single dose (acute toxicity) or after 1st of 5 daily doses (subacute toxicity).
[c] COBS of Charles River, France: Caesarean-originated, barrier-sustained.

1. Acute Toxicity and Subacute (5 Days) Toxicity

a) Method of Study

Table 22 shows the general conditions adopted in the determination of the acute toxicity and subacute toxicity (daily treatment for five days) of rubidomycin. Rubidomycin was administered in solution—either in physiological saline given intravenously or in distilled water given intraperitoneally or orally. The animals were kept either in groups of 8 or 10 (mice), or in groups of 5 (rats and guinea-pigs), or in individual cages (rabbits and dogs). They received as much as they wanted of a balanced diet suited to their species (that of the Usine d'Alimentation Rationelle, France), the dogs also receiving supplementary meat.

b) Results

The results are shown in Table 22.

α) Acute Toxicity

Administered intravenously, rubidomycin has about the same toxicity for the mouse and rat on the one hand (LD_{50} approximately 20 mg/kg for both species) and for the guinea-pig, rabbit, and dog on the other hand (LD_{50} of the order of 4 to 6 mg/kg for each species); thus its toxicity for the three latter species is about four times as great as for the mouse and rat.

In all cases the animals died, generally from the third day after administration onwards, with no particular toxic symptoms but in a state of deep stupor.

The rats, guinea-pigs, rabbits, and dogs surviving 15 days after administration of rubidomycin were sacrificed. Neither in weight nor in macroscopic appearance did the principal organs of these animals show any abnormality.

Just before they were sacrificed, these animals were subjected to a hematologic examination and to bone marrow investigation. In the rats there were no abnormalities detected; in the dogs there was slight hypochromic anemia, but only when the dose administered was 4 mg/kg intravenously; and in the guinea-pigs regenerative anemia was observed, marked when the dose was 6.2 mg/kg intravenously but very slight at lower dosages. On the other hand, the blood and marrow of the rabbits were very much more affected, as shown by a tendency to anemia of the hypochromic type at the dosage of 6.2 mg/kg intravenously, marked hyperthrombocytosis, toxic granulations in the neutrophils with or without leukopenia, and an impoverished marrow with immature granulocytic cells predominating at dosages of 6.2 and 3.1 mg/kg intravenously.

Given orally, rubidomycin was 10—15 times less toxic than given intravenously in the mouse, rat, and dog.

STERNBERG and PHILIPS [207] have described a biphasic development of toxic symptoms and lesions in the rat after a single intravenous injection of 20 mg/kg. All the animals died within a period of about a month. For the first week after the injection lesions of the lymphoid structures and the intestinal epithelium were mostly observed, with the bone marrow severely affected; some animals died at this stage. During the second week these toxic signs appeared to be less severe, but the serous membranes became affected, with ascites and pleural effusions, and renal lesions and testicular aplasia developed.

β) Subacute Toxicity

The LD_{50} doses in mice, given intravenously and intraperitoneally in daily amounts repeated for five days, were respectively about 10 and 2 mg/kg daily. In these conditions, by these two routes of administration, rubidomycin was therefore twice as toxic as it was in a single administration.

2. Long-term Toxicity (3 Months)

a) Method of Study

The study of the long-term toxicity of rubidomycin was carried out in the conditions shown in Table 23. The nature and the periodicity of the examinations carried out during the study are shown in Table 24.

b) Results

α) Trials on Rabbits

(a) Clinical Examination

At a daily dose of 2 mg/kg administered intravenously, the six rabbits died between the fourth and the tenth day of treatment. Among the animals surviving at the

Table 23. *Long-term toxicity (3 months). General conditions*

Animal	Species	Rabbit	Dog
	Race	Fauve de Bourgogne	Beagle
	No. per dose and sex	6 (males)	2 (1 male and 1 female)
Daily [a] dose [b] mg/kg i.v.		Control (vehicle only)	Control (vehicle only)
		2	
		1	1
		0.5	0.5
		0.25	0.25
Administration	Route	i.v.	
	Mode	in aqueous solution in sterile hypertonic glucose solution (30%)	
	Unit volume	2 ml/kg	1 ml/kg
Duration of treatment		3 months [c]	
No. of animals kept under observation for 6 weeks after end of treatment		2 controls and 2 on each dose [c]	0

[a] Except Saturdays and Sundays for the rabbits, except Sundays for the dogs.

[b] These doses represent respectively about 1/2, 1/4, 1/8, and 1/16 of the acute intravenous LD_{50} per kg in the rabbit or dog.

[c] The 6 rabbits given an intravenous daily dose of 2 mg/kg died within 10 days of the start of treatment. The 2 dogs given an intravenous dose of 1 mg/kg died after 12 and 13 days of treatment respectively.

Table 24. *Long-term toxicity (3 months). Examinations carried out*

Nature of examination		Periodicity of examinations (no. of animals examined per dose in bracket)	
		Rabbit	Dog
Clinical examination	General examination	Daily (all)	Daily (all)
	Weight	Weekly (all)	Weekly (all)
Hematological examination	Hemoglobin level RBC, WBC, platelet, reticulocyte count WBC formula	Before treatment, then after 1, 2, and 3 months (3 rabbits and 3 controls per dose), then 6 weeks after end of treatment (2 rabbits per category) [a]	Before treatment (twice), then weekly (all)
	Coagulation time Howell, Quick, (thrombin) Bleeding time	Not examined	
	Bone marrow	Before treatment and at end of treatment (3 rabbits of each category) [b]	Before treatment and at the end of treatment (all)
Biochemical examination	Blood — Urea	Periodicity as for circulating blood above	
	Blood — Serum electrolytes (Na, K, Ca, Cl) Glucose BSP test Transaminases (GOT and GPT) Acid and alkaline phosphatase Bilirubin Fibrinogen	Not examined	Before treatment, then after 1, 2, and 3 months of treatment (all)
	Urine — PSP test Glucose Albumin Urobilin Bile salts Bile pigments	Not examined	
Pathological examination	Macroscopic — Observation and weighing of main organs [e]	At end of treatment (4 rabbits per category) or 6 weeks after end of treatment (2 rabbits per category)	At end of treatment (all)
	Microscopic — Histological examination of main organs [d]		
	Microscopic — Special examination: histochemistry	Search for lipids in liver (in the case of observed abnormality)	Search for lipids, glycogen, and hemosiderin in liver (in case of observed abnormality)

end of the first week of treatment, only slight loss of weight was observed and, some days before death, an abnormally prolonged bleeding at the injection site, the marginal vein of the ear.

At a daily dose of 1 mg/kg administered intravenously, there was no death during the three months of treatment, but one of the two rabbits kept under observation was found dead seven days after the treatment was stopped. Weight gain in the animals treated with this dose was slightly retarded as compared with the controls, the average initial weights respectively being 2.030 kg and 2.340 kg and the average final weights being 2.735 kg and 3.470 kg. Apart from poor local tolerance at the injection site (marginal vein of the ear), the appearance and behavior of the animals remained normal during the whole of the trial.

At daily doses of 0.5 and 0.25 mg/kg administered intravenously, no death occurred during the whole of the trial and the weight gain of the animals treated was similar to that of the controls. Local vein tolerance at the injection site (again the marginal vein of the ear) was poor and little different from that observed in rabbits treated with a daily intravenous dose of 1 mg/kg.

Clinically, no abnormality was observed in the control animals. Local tolerance of the hypertonic glucose solution used as vehicle for the rubidomycin was very good.

(b) Hematological Examination

At a daily dose of 2 mg/kg administered intravenously, very clear signs of pancytopenia were observed in the circulating blood of the animals surviving at the end of the first week of treatment. They consisted of a slight tendency to anemia with a very great fall in the reticulocyte level, leukopenia, with an average of 2100 leukocytes per mm³ as against 6500 before treatment, extreme granulocytopenia, with an average of 3⁰/₀ of granulocytes, and an almost complete disappearance of platelets.

At a daily dose of 1 mg/kg administered intravenously, all that was observed was a slight overall tendency to leukopenia and a very moderate anemic tendency, which both disappeared spontaneously after six weeks of uninterrupted treatment. At the end of the third month of treatment, except for marked erythroblast activity in the circulating blood and the marrow, the hematological picture was identical with that of the controls.

At daily doses of 0.5 and 0.25 mg/kg administered intravenously, no abnormality of the blood was observed as compared with the controls.

---

^a These examinations were also carried out on all the controls and the rabbits given a dose of 1 mg/kg intravenously after 2 and 6 weeks of treatment. The platelet count was not done after 3 months of treament on the controls or on the rabbits given the dose of 0.25 mg/kg, nor on the rabbits kept under observation for 6 weeks after the cessation of treatment.

^b This examination was qualitative, not quantitative, in the rabbits. It was also carried out on the animals kept under observation for 6 weeks after the cessation of treatment.

^c Liver, kidneys, spleen, adrenals, and testicles in the rabbits and dogs. In the dogs, thyroid, heart, brain, pituitary, prostate, ovaries, uterus, and lungs as well.

^d Organs listed above for the rabbits, but in addition, for both species of animals, stomach, small intestine, colon, pancreas, bladder, lungs, thyroid, heart, striated muscle. For the dogs alone the tongue, esophagus, parathyroids, nodes, thymus, prostate, ovary and cornua of uterus and aorta were also examined.

(c) Biochemical Examination

The blood urea level remained normal in all the animals during the whole of the trial.

(d) Pathological Examination

At a daily dose of 2 mg/kg administered intravenously, hemorrhagic lesions were observed at autopsy at various sites (digestive tract, heart, lungs) [1].

At a daily dose of 1 mg/kg administered intravenously, in two rabbits out of six (one died eight days after treatment was stopped and the other was sacrificed after three months of treatment) a serofibrinous effusion was observed at autopsy in the thoracic cavity, along with a thickening of the pericardium. Histological examination of four rabbits sacrificed after three months of treatment showed considerable lesions of the myocardium—of the type of fatty necrosis or degeneration of the muscle fibres—and of the kidneys—of the type of tubular nephropathy—, as well as some hepatic and especially mediolobular steatosis. Histological examination of one of the two rabbits kept under observation after treatment was stopped and surviving after six weeks suggested that the renal and hepatic lesions were reversible; on the other hand, six weeks after the cessation of treatment the myocardial lesions left serious sequelae: scarring sclerosis with atrophy of many myocardial fibers.

At a daily dose of 0.5 mg/kg administered intravenously, the organs of all the rabbits appeared macroscopically normal. Histological examination merely showed some adrenal or renal congestion in some rabbits and, in two of the four sacrificed after three months of treatment, a fall in spermatogenesis but with continuing spermatozoid production in one or several testicular lobules. Histological examination of the two rabbits kept under observation after cessation of treatment revealed only some areas of slight interstitial fibrosis of the myocardium.

At a daily dose of 0.25 mg/kg administered intravenously, testicular atrophy was seen at autopsy in one of the four rabbits sacrificed after three months of treatment. On histological examination, some slight signs of hepatic involvement of the steatosis type were found in one rabbit sacrificed after three months of treatment and, in one kept under observation after cessation of treatment, portal fibrosis and slight interstitial sclerosis in the right ventricle. Also in one rabbit histological examination showed total absence of spermatogenesis in one testis, the other being normal.

At autopsy, testicular atrophy was observed in two of six control rabbits examined (in one sacrificed six weeks after treatment was stopped a testicle was ectopic). In the rabbit sacrificed after three months of treatment, histological examination revealed cessation of spermatogenesis at the spermatocyte stage, absence of spermatozoids in the lumen of the seminiferous tubules and the epididymis, and hyperplasia of the Leydig cells. One of the two testicles showed almost complete Sertoli cell regression.

(e) Conclusion

In the rabbit at a daily dose of 2 mg/kg administered intravenously, rubidomycin showed a marked tendency to affect the hemopoietic system, with pancytopenia of the circulating blood, and the animals died after four to ten days of treatment.

---

[1] Since these animals were found dead after four to ten days of treatment in a state of advanced autolysis, no histological examination was possible.

After three months of treatment at a daily dose of 1 mg/kg administered intravenously there was no mortality, but tolerance was low. There were transitory blood disturbances (a tendency to anemia and leukopenia), hepatic steatosis, renal and myocardial lesions, and a slight slowing down of growth. The renal and myocardial lesions—of the type causing muscle fiber necrosis or degeneration—are the most noteworthy because they seemed to have serious sequelae.

When administered intravenously to rabbits, the maximum daily tolerated dose of rubidomycin in a three-month course of treatment is of the order of 0.25—0.5 mg/kg. With such a dose the possibility of testicular involvement cannot be eliminated although, oddly enough, with a dose of 1 mg/kg it was not observed, although analogous testicular conditions were also observed in one control animal. The changes seem, however, to be reversible, since they were not found in the animals sacrificed six weeks after the treatment was stopped.

β) Trials on Dogs
(a) Clinical Examination

At a daily dose of 1 mg/kg administered intravenously, the two dogs died after 12 and 13 days of treatment respectively, having towards the end of the first week shown a tendency to anorexia and a fairly marked disinclination to movement. Two to three days before their death, the appearance of some petechiae was observed on the inner aspect of their jowls and on their gums.

At a daily dose of 0.5 mg/kg administered intravenously, the two dogs progressively lost weight for the first five weeks of treatment (15—20% of their initial weight by the end of the fifth week). From the end of the fourth week on they showed marked anorexia and fairly marked disinclination to movement and presented small buccal lesions of the necrotic type. All their disturbances were accompanied by clearly marked abnormalities of the blood. Their general state then progressively improved and at the end of treatment, although the treatment was uninterrupted throughout, they had become normal in appearance and behavior and their weight had returned to the initial value.

At a daily dose of 0.25 mg/kg administered intravenously, the two dogs showed no clinical abnormality.

At all the daily dosages studied, local toleration at the injection sites was good, except for one dog treated with 0.5 mg/kg, in which the accidental injection of a small amount of the solution into the perivascular tissues on the 28th day of treatment caused considerable tissue damage—abscess formation, then necrosis. After local treatment with antibiotics, however, the tissue healed completely within a month.

(b) Hematological Examination

At a daily dose of 1 mg/kg administered intravenously, the two dogs that died after 12 and 13 days of treatment respectively showed serious disturbances of the blood picture: leukopenia and granulocytopenia to an intense degree (800 and 600 leukocytes per mm^3 of blood), thrombocytopenia (moderate in one dog but very great in the other, with a fall of 90% in the number of platelets), and a very marked increase in the bleeding time (more than eight minutes as against two to three before treatment). In both dogs the erythrocyte series remained normal except for the almost complete disappearance of reticulocytes.

At a daily dose of 0.5 mg/kg administered intravenously, only one of the dogs had—at the end of the fourth week—an episode of leukopenia with granulocytopenia (2800 leukocytes per mm³ and 12% granulocytes as against 15 100 and 58% before treatment), accompanied by marked thrombocytopenia (a fall of 85% in the number of platelets), without any very obvious increase in bleeding time. By the end of the fifth week of treatment the leukocyte and platelet values had returned to normal, and they remained so till the end of the trial. At the end of the fourth week a moderate normochromic anemic tendency, very weakly regenerative, was seen in this same animal, and it continued practically unchanged throughout the treatment, with a fall in the erythrocyte level varying between 15% and 35%.

Disturbances of the blood were more marked in the other dog treated with a daily dose of 0.5 mg/kg administered intravenously. During the first three weeks of treatment a slight normochromic anemic tendency was observed. Then the intensity of the anemia increased: at the twelfth week erythrocyte levels were about 60% less than they had been initially and no clear sign of regeneration was observed. From the fourth to the tenth week there was a moderate leukopenia (3000—3400 leukocytes per mm³ as against 7600 before treatment), which later disappeared. A marked thrombocytopenia began at the end of the fourth week, accompanied by an increase in the bleeding time to more than 10 minutes as opposed to about two before treatment. The severity of the thrombocytopenia—a fall of from 70% to 90% in the number of platelets—continued practically unchanged till the penultimate week of treatment, but slackened off to about 50% of the initial level at the end of the last week. From the end of the eleventh week on, however, in spite of the still very marked numerical deficit in the number of platelets, the bleeding time returned to normal limits.

Examination of the bone marrow of the two dogs treated with a daily intravenous dose of 0.5 mg/kg, carried out after three months of treatment, only showed rejuvenation of the granulocytic and erythrocytic series with no change in the ratio of granulocyte to erythrocyte series.

At a daily dose of 0.25 mg/kg administered intravenously, the various hematologic values remained within normal limits. At the end of the treatment, the myelogram showed no abnormality.

(c) Biochemical Examination

At a daily dose of 0.5 mg/kg administered intravenously, apart from slight transient rises in urobilin and urinary bile salts in the two dogs, and albuminuria and glycosuria in one dog, an increase in plasma fibrinogen was observed after one month of treatment (fibrin levels of 6.7 and 14.7 g/liter respectively), with a return to normal (3—5 g/liter) after two and three months of treatment.

At a daily dose of 0.25 mg/kg administered intravenously, the only abnormalities noted were slight albuminuria after one, two, and three months of treatment in the two dogs and a slight transient increase after one month of urobilin and especially bile salts in the urine of one dog.

(d) Pathological Examination

At a daily dose of 1 mg/kg administered intravenously, the autopsy showed that there were lesions of an essentially hemorrhagic nature in the gastric mucosa, intestine, and spleen [2].

At the daily dose of 0.5 mg/kg administered intravenously, the autopsy showed, in the male dog, testicular atrophy only. Histological examination merely showed total aspermatogenesis in this dog, associated with hyperplasia of the Leydig cells. In the other dog there were some hepatic changes: considerable centrolobular stasis with atrophy of the strands of the hepatic cell and degeneration of certain cells in contact with the centrolobular veins, Küpffer's cell reaction with initiation of collagen formation in the reticulum, the presence of small lipid drops along the axis of the liver cells, and the accumulation of hemosiderin in some of the liver cells, in the areas of stasis, and in some Küpffer's cells. The hemosiderin deposits, like the cell changes, can be considered a consequence of the stasis and show that it had lasted for some time; there was therefore no question of its being a terminal stasis.

At the daily dose of 0.25 mg/kg administered intravenously, testicular atrophy was noted at autopsy in the male dog. Histological examination revealed total aspermatogenesis with hyperplasia of the Leydig cells. This lesion, it would appear, was irreversible, since close examination showed no spermatogonia in the seminiferous tubules, which were therefore totally aplastic. In the same dog congestive areas were observable in the mucosa of the digestive tract, the lung, the liver, and the spleen, with hemosiderin deposits in the reticular cells of the spleen and Küpffer's cell of the liver. No abnormality was found in the other dog—a bitch—treated with this dose.

(e) Conclusion

In the dog at a daily dose of 1 mg/kg administered intravenously, rubidomycin particularly affected the blood (aleukia hemorrhagica), the animals dying after 12 or so days of treatment.

After three months of treatment at a daily dose of 0.5 mg/kg administered intravenously, the mortality was nil but the tolerance was low, as indicated by weight loss (transient), blood disorders—anemia, very marked but transient leukopenia, thrombocytopenia and, possibly, prolongation of the bleeding time—, and total aplasia of the spermatocyte series in the seminiferous tubules with complete aspermatogenesis.

When administered intravenously to dogs, the maximum daily tolerated dose of rubidomycin in a three-month course of treatment was approximately 0.25 mg/kg. Even here an exception must be made for testicular changes with complete aspermatogenesis, which are probably irreversible.

3. Teratogenic Action

The possible teratogenic action of rubidomycin was studied in the chick embryo, the mouse, and the rabbit.

a) Chick Embryo

Fertilized eggs of Leghorn hens of the same brood were given an injection into the yolk-sac on the third day of incubation of 25 or 50 µg of rubidomycin per egg

---

[2] Since the animals were found dead after 12 and 13 days of treatment respectively in a state of advanced autolysis, no histological examination of the chief organs was possible.

contained in 0.10 ml of distilled water. Some eggs were left without an injection as complete controls, others were given an injection of 0.10 ml of distilled water into the yolk-sac (distilled-water controls). The eggs were candled daily, and from the eighth day on the dead embryos were subjected to anatomical examination. The embryos surviving to the15th day were all taken, weighed, and examined for abnormalities.

The results obtained are shown in Table 25.

Table 25. *Teratogenic effect on chick embryo*

Batch	No. of em-bryonated eggs	Embryos dead during trial %	Abnormalities observed %	Average weight of embryo (g)
Complete controls	53	7	2	12.4 ± 0.1
Distilled-water controls	53	13	8	11.0 ± 0.2
Rubidomycin 50 μg/egg	55	40	2	10.9 ± 0.1
Rubidomycin 25 μg/egg	55	29	2	11.7 ± 0.3

Even at doses that were toxic to the embryos (25 μg and 50 μg per egg), rubido-mycin had no teratogenic action on the chick embryo. The few abnormalities observed were those we are accustomed to seeing during such experiments; whatever the batch, they were cephalic abnormalities or visceral hernias.

b) Mice

Two-month-old G.S. female mice reared by us, in batches of 20, were treated with a daily dose of 1.25 mg/kg of rubidomycin subcutaneously for a week before mating. They were placed in the presence of males—one female to one male—for 20 days, during which period only the females were given rubidomycin. They were then

Table 26. *Teratogenic effect on mouse*

		Controls	Rubidomycin 1.5 mg/kg s.c.
No. of pregnant mice		20/20 (100%)	20/20 (100%)
Total no. of young mice		125	132
Average no. of young mice per female		6.2	6.6
Cannibalism		0	8
Young mice dead during 1st month		0	1
Total no. of mice aged 1 month		125 { 61 males 64 females	123 { 62 males 61 females
Abnormality observed		0	0
Average weight (g) of young at birth		1.72	1.66
Average weight increase (g) of young during 30 days	8 days	3.46	2.89
	16 days	6.80	5.91
	24 days	13.29	11.24
	30 days	18.70	16.48
Average weight (g) of mouse aged 1 month		20.42	18.24

separated from the males and rubidomycin was continued till they had given birth. At birth the young were counted, examined, weighed, and observed up to the age of one month. At that age all the animals were sacrificed and given a post-mortem examination for abnormalities.

The results obtained are shown in Table 26.

The table shows that, despite the prolonged treatment, rubidomycin caused no disturbances in gestation and had no teratogenic effect. In comparison with the control batch of mice, however, the gain in growth of the batch whose mothers had been treated was slightly retarded.

c) Rabbits

Virgin "Fauve de Bourgogne" rabbits from the same producer, weighing between 3.2 and 4.2 kg, were treated with rubidomycin from the day of mating for 15 days, the daily dose, administered intravenously, being 0.25 to 0.05 mg/kg according to the batch. On the 18th day, abdominal palpation enabled the extent of the pregnancy to be gauged. The rabbits were sacrificed on the 29th day after mating and the uterus was examined for indications of the number of fetal resorptions. The young were removed, counted, weighed, and autopsied, and in certain cases the skeleton was colored with alizarin.

The results obtained are shown in Table 27. Intravenous doses of 0.25 mg/kg and 0.10 mg/kg clearly caused abortion (100% and 66% respectively). The dose of 0.05 mg/kg was better tolerated, with 25% of abortions. The 11 litters produced 97 young, with 8.8 young per litter on the average. Four abnormalities were observed, i. e., 4.1%: a percentage slightly higher than that recorded during the examination of 400 control fetuses (abnormalities 1.5%). The percentage of fetal resorptions was similarly slightly higher than observed as a rule: 11% as against 8%. These results, not very different from the usual, suggest that rubidomycin, in spite of its high toxicity for the embryo, has no teratogenic effect on the rabbit.

d) Conclusion

Rubidomycin has no teratogenic effect on the chick embryo, mouse, or rabbit.

4. Carcinogenic Effect

It is well known that substances with antitumoral activity may, paradoxically, exert a carcinogenic effect. For this reason we considered it of particular interest to study the carcinogenic effect of rubidomycin.

a) Material and Method

α) Subcutaneous Administration

A weekly subcutaneous injection of 1.25 mg/kg of rubidomycin was given to XVII/Rho mice, aged five weeks at the beginning of the test, for a period of 12 weeks. In all, a dose of 15 mg/kg was administered. The animals were in the following batches:

> treated batch: 20 males and 20 females
> control batch: 20 males and 20 females.

Table 27. *Teratogenic effect on rabbit*

Rubidomycin mg/kg i.v.	No. of female rabbits treated	No. of abortions	Litters		No. of fetal resorptions	No. living	Average weight of young (g)	Observations
			No.	Serial no.				
0.25	6	6						
0.10	15	10	5	1	1	7	37.5	
				2	0	8	43.8	1 parieto-occipital cranioschisis
				3	0	11	32.4	
				4	4	9	24.4	2 umbilical hernias
				5	0	7	41.6	
0.05	8	2	6	1	0	10	38.5	
				2	3	7	34.3	
				3	1	10	37.0	1 rachischisis
				4	1	9	42.2	
				5	2	10	40.0	
				6	0	9	38.1	

The control animals received the same volume of solvent as the treated animals, viz. 25 ml/kg of distilled water subcutaneously.

β) Oral and Intraperitoneal Administration

A weekly oral dose of 12.5 mg/kg of rubidomycin was given to $C_{57}Bl/Rho$ mice aged about two months at the start of the trial, for 22 weeks. In all, a dose of 275 mg/kg was administered.

Other $C_{57}Bl/Rho$ mice of the same age were given a weekly intraperitoneal injection of 0.625 mg/kg making a total dose of 11.87 mg/kg.

The distribution of the animals was the usual: 20 males and 20 females in each batch.

b) Results

The mice of the various batches were observed daily and weighed monthly. Every sick mouse was killed, and a full autopsy and a histological examination of the organs were carried out. At the end of 22 months of observation, the surviving mice were sacrificed and examined as in the case of those which had died. The results obtained are shown in Table 28.

Table 28. *Carcinogenic effect on mouse (duration of observation 22 months)*

Batch			Pulmonary adenoma	Hepatoma	Reticulosarcoma Leukosarcomatosis Leukosis	Local fibrosarcoma	Various benign tumors
XVII/Rho mice	Controls	20 males	11	4	0	0	0
		20 females	15	1	0	0	0
	Rubidomycin s. c.	20 males	10	1	0	5	0
		20 females	11	0	0	5	2
$C_{57}Bl/Rho$ mice	Controls	20 males	2	4	4	0	3
		20 females	4	1	8	0	4
	Rubidomycin oral	20 males	0	2	1	0	5
		20 females	1	1	3	0	3
	Rubidomycin i. p.	20 males	1	1	0	0	1
		20 females	0	0	0	0	3

α) Subcutaneous Administration

When administered subcutaneously, rubidomycin caused fibrosarcomas to appear at the injection site. Of the animals treated 25% had sarcomas, while the controls had none.

β) Oral and intraperitoneal Administration

The frequency of tumors in the treated animals was little different from that in the controls. There were, however, fewer reticulosarcomas, leukosarcomatoses, and leukoses in the animals treated orally and fewer especially in those treated by intraperitoneal injection. It must, however, be remembered that these conditions appeared late, towards to 16th—17th month in the controls, at a time when the numbers of mice in the treated batches had gone down. To take that situation into account, the position from the 16th month on should be assessed as follows:

Controls: 35 mice, of which 12 had reticulosarcomas or leukoses (34%).

Oral rubidomycin: 19 mice, of which 4 had reticulosarcomas or leukoses (21%).

Intraperitoneal rubidomycin: 9 mice, of which none had reticulosarcomas or leukoses (0%).

Statistical analysis of these figures shows that the difference observed in relation to the controls is not significant, either for the batch treated intraperitoneally ($chi^2 = 3.01$; $0.10 > p > 0.05$) or for the batch treated orally ($chi^2 = 0.49$; $p = 0.50$).

c) Conclusion

Rubidomycin, when injected subcutaneously into mice, causes fibrosarcomas to develop at the injection site. When administered orally or intraperitoneally, it has no carcinogenic effect.

Part II — Clinical and Therapeutic Study

Chapter 6

Posology

General Management of Treatment. Symptomatic Treatment

1. Dosage with Rubidomycin

Rubidomycin can be used alone or in association with other drugs. The dose varies according to the disease and according to the stage of the disease, the details being given elsewhere for each of the main therapeutic indications. We shall confine ourselves here to some general remarks.

2. Presentation of Rubidomycin

Rubidomycin (13.057 RP) is presented in an injectable lyophilized vial containing a quantity equivalent to 20 mg of base in the form of the hydrochloride, in accordance with the following unit formula:

rubidomycin (13.057 RP) hydrochloride	21.6 mg
(i. e., 20 mg of base)	
anhydrous monosodium phosphate	72.6 mg
anhydrous disodium phosphate	3.96 mg

Daunomycin (1762 F) is presented in an injectable vial containing a quantity equivalent to 20 mg of base:

daunomycin hydrochloride	20 mg
mannitol	100 mg

3. Route of Administration

As things are at present, the drug should be administered strictly only by the intravenous route. After it is dissolved in 10 ml of physiological solution, it is passed into the tubing of an intravenous perfusion of 100 ml of physiological solution. The perfusion is administered very rapidly after the drug, so as to avoid local stasis of rubidomycin to the greatest possible extent.

4. Dosage

Injection dose. The injection dose varies from 0.5 mg/kg to 3 mg/kg, i. e., from 15 to 100 mg/m².

Daily injections are given in particular in vigorous treatment of acute leukemias and when rubidomycin is used alone. A quick effect is necessary, and the risk of a sudden severe aplasia is taken into account, either because the leukemia is an immediate threat—as with acute leukemias with promyelocytes or acute myeloblastic leukemias—or because no combination of drugs is possible since the patients have been treated previously and have become resistant to other drugs.

Injections are given *every 2—3 days when* marrow aplasia is to be avoided. It is then possible to spread the treatment over a period of time and the effect of the rubidomycin can be regularly watched by a hemogram before each new injection. This is the treatment for acute myeloblastic leukemias when several drugs are used in combination with rubidomycin, and for chronic myeloid leukemia, sarcomas, Hodgkin's disease, and carcinomas.

Injections are given *once a week* in particular in therapeutic combinations, the most classic case at present being the combination used in the treatment of acute lymphoblastic leukemias and certain lymphosarcomas, i. e., vincristine, rubidomycin, and prednisone.

Total dose in a series of injections. The total dose given in a series of injections varies from 3 mg/kg to 22 mg/kg (90—660 mg/m²).

Total cumulative dose. The total dose administered is strictly limited by the fear of cardiac complications and the dangers of the accumulation of the drug. The total of 25 mg/kg (750 mg/m²) should in principle not be exceeded. In certain cases, each discussed on its merits, total doses of 25—30 mg/kg (900 mg/m²) may be prescribed, particularly for patients who have become resistant to all drug treatment and who may, in spite of the risk to the heart, achieve a remission by a supreme effort.

5. General Management of Treatment. Precautions

a) Symptomatic Treatment

Because of the powerful effects of rubidomycin, it cannot be administered unless stringent precautions are taken.

b) Place of Treatment

The initial treatment can take place only in a hospital, preferably in a specialized unit.

At later stages and for reinduction treatment, short periods of stay in hospital may be feasible, or the system of day hospital may be employed. When the first reinductions are well tolerated, it may be possible afterwards to carry them out as consultations, with the patients ambulant.

c) Special Treatment of Hyperleukocytic Forms and Very Active Multiple Drug Therapy

The following three measures should be taken:

1. Immediate administration of Allopurinol in a dose of 600—800 mg daily (10—15 mg/kg daily taken in three divided doses after meals).

2. Frequent drinks—of the order of 3—4 liters every 24 hours for an adult—, half of which should be alkaline, particularly sodium bicarbonate. For the infant and child, an intravenous transfusion of isotonic glucose solution should be given for the first 4—5 days of treatment (1000 ml daily per 20 kg body weight). For the child as well as for the adult, the blood urea and uric acid levels should be checked 3—4 times during the first week, and the blood and urinary electrolytes should be determined. Potassium should not be added to oral corticosteroids till the third or fourth day of treatment according to the results of the blood electrolyte determination.

3. A short delay before rubidomycin treatment is begun, Allopurinol and plenty to drink being recommended for 2—3 days before it starts. The latter treatment should be continued with a reduced dose of Allopurinol (300—400 mg daily) in balance with the uric acid level until the danger of hyperuricemia and hyperuricosuria is past.

These measures obviate the danger of serious accidents—hyperuricemia, hyperkalemia, hyperazotemia, and sometimes even anuria, caused by the precipitation of uric acid crystals in the ureters. The very rapid destruction of an extremely large number of leukemic cells may be responsible for accidents of this kind.

6. Symptomatic Treatment

It seemed to us very important that, in using a very active drug like rubidomycin, we should accept the risk of a sudden myeloid aplasia so as to obtain the total remission of an acute leukemia when it abated. Such a risk can be taken on the sole condition that the complications and duration of the aplasia should be able to be reduced to the maximum possible degree. By the use of transfusions of platelets, the rational employment of antibiotics, and transfusions of white blood cells, aplasia complicated by hemorrhage and/or infection can be controlled far more frequently than it used to be.

Another prerequisite is that the patient should be isolated in conditions that are as aseptic as possible.

a) Isolation

The isolation and care of the patient in germ-free rooms gives additional security and enables doses of rubidomycin to be given that cause aplasia of the bone marrow while diminishing to the greatest degree possible the risks to the patient's life [174].

Employment of the sterile plastic bubble is probably the best solution at present, but we have not as yet been able to use it routinely. In most cases our patients have been isolated in conditions that were as aseptic as possible.

b) Platelet Transfusion [92, 209]

Platelet transfusions are systematically indicated for patients with leukemia receiving intensive treatment with rubidomycin, to enable them to surmount the stage of bone marrow aplasia that is inevitable during treatment. The platelets transfused should be viable, effectively ensure hemostasis, and be injected in adequate quantities.

They should be transfused repeatedly for more or less lengthy periods—ranging from several days to several weeks. An important problem that the transfusions raise is that of the progressive immunization of the recipient, since immunization brings with it excessive destruction of the platelets and ineffectiveness of the transfusions.

With our present technical knowledge we cannot keep platelets for more than a few hours. Whatever the different techniques used, the lapse of time between taking the platelets and the transfusion should not be more than 6 hours.

The authorities all seem to agree at present that the best transfusion results are obtained if the platelets are taken in DCA solution, hyperacidified to pH 6.5. This acidification of the medium is an essential condition for preventing spontaneous clumping of the platelets. The transfusion results with platelets in DCA solution hyperacidified to pH 6.5 are twice as good as those with platelets in normal DCA, and at the time of the transfusion the spleen sequesters the platelets hardly at all.

Fresh blood collected less than 6 hours previously in a vessel with wettable walls provides few platelets in comparison with what is needed for hemostasis. A transfusion of such blood is to be reserved for cases in which thrombocytopenia is associated with some degree of anemia.

Plasma rich in platelets (PRP), obtained by centrifugation of blood at slow speed (1200 revolutions per minute for 10 minutes at 8—12° C) to get rid of the red blood cells exposes the patient to the danger of accidents from plasma overloading, given the total volume that must be injected into the patient to compensate for the platelet deficit.

The best method is to inject platelet concentrations containing considerable quantities of platelets in small volume. Platelet concentrations like this are obtained by centrifuging PRP at 8—12° C at high speed (3600 revolutions per minute for 30 minutes). The platelet sediment at the bottom is re-suspended in 20—40 ml of plasma; the rest of the impoverished plasma can be used for other purposes, e. g., for the treatment of hemophilia. With excess citrate in the medium the platelets can easily be replaced in suspension in a very small amount of plasma. Only the second vessel containing the platelet-rich plasma is given excess citrate, so that the sediment of red blood cells can be used at a neutral pH.

The amounts of platelets to be injected to obtain the right degree of hemostasis in patients with aplasia are considerable. They should be calculated in terms of the weight and height evaluated in m^2 of body surface. On the basis of one unit of platelets, i. e., the platelet concentration obtained from a 500 ml bottle of blood with 200 000 platelets per mm^3, containing at most 1×10^{11} platelets and being sufficient to increase the platelets from 12 000 to 14 000 per mm^3 in a subject with a body surface of 1 m^2 (a child of 30 kg, say), that subject must be given 4 units of platelets to raise the platelet count by 50 000 per mm^3. To achieve the same result in an adult weighing 60 kg, 8 units of platelets must be injected.

This calculation is valid only when the platelets are fully utilized, when the patient is neither immunized against them nor infected. It is known that when platelet transfusions are given repeatedly they become less and less effective, raising the problem of the receiver's immunization.

c) Bacterial Infections [90, 219]

Bacterial infections, especially infections with enterobacteria or staphylococci, are the main cause of death in patients with acute leukemia who become aplastic. For such patients, therefore, antibiotic treatment is of crucial importance. It should, however, have already been discussed with the patient, since the considerable chance that he will have aplasia on the treatment is well known; in other words, the question of preventive antibiotic treatment arises at the outset of treatment for leukemia.

α) Antibiotic Treatment before the Stage of Aplasia

There are three possibilities before the stage of aplasia:
1. The patient may not be feverish and may apparently not be infected.
2. The patient may be feverish and may be obviously infected.
3. The patient may be feverish, but his fever cannot with certainty be related to bacterial infection.

If the patient is not feverish and apparently not infected, there is no justification for the use of antibiotics, since it risks selecting out a micro-organism resistant to the antibiotics used and so may, afterwards, when the aplastic phase is abating, be responsible for a severe infection and even a septicemia refractory to antibiotic treatment.

If the patient is feverish and obviously infected—with, for example, a buccopharyngeal infection or a pyodermatitis—immediate antibiotic treatment is in most cases necessary, though theoretically undesirable until the microorganism responsible for the infection has been identified.

While awaiting laboratory identification of the micro-organism and its susceptibility to antibiotics, a combination of the type penicillin or cephalosporin plus streptomycin or kanamycin may be used. These antibiotics have the double advantage of acting in most cases synergetically and of having a very wide spectrum, acting on both the gram-positive and gram-negative organisms most commonly encountered in infections in leukemia: enterobecteria and *Staphylococcus aureus*.

Lastly, if the patient is feverish but it is not certain whether he has a bacterial infection, antibiotic treatment should not be started until the laboratory has reported its results on the basis of repeated blood, urine, and stool specimens and a throat swab.

Because of the seriousness of their condition, it is often necessary to adopt the same approach to these patients as when they are obviously infected, since it is difficult not to attribute their feverishness to latent infection.

β) Antibiotic Treatment during the Stage of Aplasia

The employment of antibiotics during the stage of aplasia also depends on the condition of the patient.

If the patient is not feverish and not infected, there is no indication for preventive antibiotic treatment, for the same reasons as when he is not in the aplastic state and is afebrile: there is a risk that treatment would select out a micro-organism resistant to the antibiotics employed, which might possibly be responsible later for a serious infection or for a septicemia refractory to treatment.

If the patient is feverish, immediate employment in high doses of a synergetic combination of broad-spectrum antibiotics, such as penicillin or cephalosporin plus kanamycin or streptomycin, often proves necessary before the laboratory has produced its results.

In many cases the patients concerned have become septicemic and the question of time is crucial.

Search for the organism responsible for the infection should be carried out as soon as possible on the basis of the repeated taking of specimens of blood, urine, stools and throat swabs. Blood for culture should be taken whenever the temperature reaches a peak, since this indicates the discharge of bacteria into the blood stream.

For blood culture two main techniques can be used: a liquid medium or Reilly's soft agar.

Throat swabs are also of much interest. When associated with and carried out at the same time as the taking of blood, they often make it possible to isolate the organism responsible for the infection faster and more reliably than by blood culture. Isolation of the same organism by both these methods furnishes an additional argument in favour of the specific bacterial origin of the infection.

γ) Minimum Inhibiting Concentration

At the best, isolation and identification of an organism and an antibiotic spectrum indicating how it will behave towards the various antibiotics will take about 48 hours in the laboratory. Though this lapse of time is considerable, it does provide accurate information on the amounts of antibiotics needed to stop the multiplication of the infecting organism in the patient's body. The minimum inhibiting values or concentrations are given in micrograms per milliliter. For the organism to be sensitive to an antibiotic, the minimum inhibiting concentrations of that antibiotic must be less than or at worst equal to the antibiotic concentrations obtained in a patient's blood after non-toxic doses have been administered to him.

δ) Choice of Antibiotic. Concept of Bactericidal Power

It often happens that antibiotics administered at doses indicated by laboratory results to be inhibitory or bacteriostatic are without obvious effect.

This may be due to the fact that there is a focus of infection into which the antibiotic does not penetrate or hardly penetrates because the serous membranes have low permeability or because the antibiotic employed diffuses poorly in the tissues.

If this is the case, the antibiotics that diffuse best must be chosen among those that are active—such as penicillin, chloramphenicol, terramycin, erythromycin, or streptomycin—, and care must be taken to avoid antibiotics that diffuse poorly or not at all, like polymycin or colimycin.

Another possibility is that the level of free antibiotic in the blood of the patient under treatment is very low, because most of it is fixed by the serum proteins.

In this case bactericidal doses of antibiotics are needed. The amount of antibiotic needed to achieve bactericidal power can be assessed by the laboratory, but here again a delay of 48 hours will occur. For rapid action while awaiting the laboratory results, provided the toxicity of the antibiotic is not a limitation, a dose five to six times that of the bacteriostatic dose can be administered, this generally corresponding to the bactericidal dose.

ε) Antibiotic Combinations

In many cases it is possible by combining two or more antibiotics to obtain a bactericidal effect without using the bactericidal dose of each of the antibiotics employed. This synergism, which is still poorly elucidated from the theoretical point of view, is due to the ability of certain antibiotics to enhance to a very great extent the action of another antibiotic. The effect is unfortunately not constant; the combination of two antibiotics may be merely the effect of each added to the other, and at times one is antagonistic to the other.

The laboratory can report on whether the antibiotics used against an infecting micro-organism are synergistic, additive, or antagonistic in effect. But it takes at least 48 hours for the report, and in practice it is therefore preferable to rely on experimental combinations.

Of such combinations, experience shows that the one most often synergistic is the combination of penicillin or cephalosporin with streptomycin, kanamycin, or gentamycin.

On the other hand, a combination of penicillin or cephalosporin with such antibiotics as chloramphemicol or the tetracyclines is mostly antagonistic. All other combinations appear to be mostly additive or synergistic.

ζ) Problems of Fungal Infections

Fungal infections are very serious and can be observed when aplasia begins to abate, though much less commonly than bacterial infections. The usual antibiotics are of little effect against them, only amphotericin B having some activity, marred unfortunately by its very great toxicity and the need therefore to use it with extreme caution. Nystatin is routinely used to prevent *Candida* infections from becoming generalized from their origin in the buccal cavity.

η) Duration of Antibiotic Treatment

Antibiotic treatment in cases of leukemic aplasia must be prolonged, even more than in cases with infection but no aplasia, and continue for at least 10 days beyond the cessation of the aplastic stage. To stop treatment too early would be to risk a relapse, which would be the harder to treat because the primary infection is in most cases partially refractory to treatment.

In conclusion, while with an appropriate antibiotic a number of patients with aplasia and a severe bacterial infection can be cured, its effectiveness is unfortunately often reduced in comparison with that obtained with non-leukemic patients affected by a severe bacterial infection because of the absence of white blood cells in the patients with aplasia.

For patients with aplasia, therefore, it is essential to combine with any form of antibiotic treatment transfusions of white blood cells, especially of the granulocyte series.

d) Leukocyte Transfusions from Donors with Chronic Myeloid Leukemia at the Myelocytic Stage [94, 97, 151, 195]

The grave complications of agranulocytosis, infections and necrosis, are in most cases resistant to antibiotic treatment. Even when administered in full knowledge of the bacteria concerned in the patient's infection, such treatment has hardly any effect on the mortality of the more severe forms.

To use donors with chronic myeloid leukemia for leukocyte transfusions has two advantages:

1. If the donor has 200 000 granulocytes per mm³, a single 500 ml bottle of his blood contains the 10^{11} white blood cells needed.

2. The younger cells can divide and mature to such an extent that a few days after the transfusion the number of circulating polymorphonuclears can be seen to rise. Various methods—study of the nuclear appendages of the polymorphonuclears or the gonosomes of the marrow mitoses if donor and recipient are of different sex, or search for the Philadelphia chromosome in marrow cells if they are of the same sex— may make it possible to demonstrate the role of the donor cells and show the reality of a temporary graft. Indeed, an erythroblastic graft of long duration has been observed.

Donors are chosen from among patients with chronic myeloid leukemia in the myelocytic stage who have at least 100 000 white cells per mm³ of blood. The presence of of hemolysin is looked for if the donor has Group 0 blood (is a universal donor).

We have on occasion, when there was AB0 and Rh compatibility, transfused whole blood collected on heparin or DCA, the donor in his turn receiving a transfusion of whole blood or packed cells from normal subjects. In other cases, plasma rich in leukocytes (and platelets) was separated out, either in plastic bags or in silicon glass vials, by spontaneous sedimentation at 37° C, by gentle centrifugation, or by the addition of Plasmagel, the red blood cells being returned to the donor.

The aplasia in the leukemic patients given white blood cells was very severe. The white blood cells in the peripheral blood were not more than 400 per mm³ and the polymorphonuclears were often absent.

In 17 cases the aplasia was total, with no blasts. In three cases the marrow was very hypocellular but still contained some blast cells (less than 20%). In the remaining four cases the marrow, though hypocellular, was still clearly blastic. In spite of vigorous antibiotic treatment, all the patients but one had a high temperature and ten showed a positive blood culture. Several patients with successive aplastic episodes were treated in this way (Table 29).

On the whole, transfusions with 7×10^9 to 2.5×10^{11} leukocytes per m² were given with very satisfying results (Table 30). By the next day a very dignificant increase was obtained in the number of leukocytes circulating in the peripheral blood after most of the transfusions: 11 times from 1000 to 2000, 7 times from 2000 to 5000, 7 times to more than 5000. In only five cases was scarcely any change seen in the blood.

In sum, the percentage of injected white blood cells found the day after the transfusion in the circulation varied from 0 to 12% (average 5.1%; median 5%).

This increase in the level of white blood cells corresponded with very favorable results. All the children responded in 24—48 hours after the transfusion by a fall in temperature and organisms were no longer found on blood culture. The results were a little less outstanding with the adults; in 9 out of 14 there was a fall in temperature, and in 2 out of 3 a negative result on blood culture. With our aplastic cases we had no serious fungal infections; if there are any, the result to be expected from the trans-

Table 29. *Leukemic aplasia treated with leukocyte transfusions from patients with chronic myeloid leukemia in the myelocytic phase (aplasia occurring at the end of treatment with rubidomycin)*

	Cell type	No. of patients	No. of aplastic episodes treated	Type of aplasia No blast cells	Blast cells	Temp $\geqq 39°$ C	Positive blood culture	
Children	Lympho-blastic	6	6	5	1	6	5	*Staphylococcus* 2 *Pseudomonas aeruginosa* .... 1 *B. coli* .... 1 *Proteus* 1
	Myelo-blastic	4	6	3	1	3	2	*Staphylococcus* 1 *Pseudomonas aeruginosa* 1
Adults	Lympho-blastic	3	4	2	1 [a]	3	1	*Pseudomonas aeruginosa* [b] ... 1
	Myelo-blastic	11	14	7	4 (2 [a])	11	2	*Staphylococcus* 2
TOTAL		24	30	17	7 (3 [a])	23	10	

[a] Slight marrow blastosis (less than 20%).
[b] With a pulmonary focus of *Pseudomonas aeruginosa* infection.

Table 30. *Results obtained from leukocyte transfusions in patients with aplasia (treated with rubidomycin for acute leukemia)*

Patients	Cell type	No. of patients	Increase in WBC > 1000	Regression of blastosis after WBC transfusion	Blood culture rendered negative	Disappearance of fever
Children	Lymphoblastic	6	6	1/1	5/5	6/6
	Myeloblastic	4	4	1/1	2/2	3/3
Adults	Lymphoblastic	3	2	1/1 ([a])	0/1	1/3
	Myeloblastic	11	8	3 (2 [a])	2/2	8/11
TOTAL		24	20	6 (3 [a])	9/10	18/23

[a] Slight marrow blastosis (less than 20%).

fusions should not perhaps be expected to be favorable for, once the infections have spread, the factors coming into play would be the phenomena of delayed immunity particularly, the polymorphonuclears playing a protective role only at the first stage of the process by preventing the fungi from penetrating deeply. Even when leuko-cytosis declined on the second or third days, clinical improvement continued in most cases. As a result at times of a reinvasion of the marrow by blast cells, chemotherapy had to be resumed, and this occasionally led to a second aplastic episode and made another transfusion of white blood cells necessary—with generally the same results as the first time.

It may be added that, in three out of the four cases of severe marrow blastosis, the condition improved markedly. This was also observed in the three patients in whom the marrow blast cells were fewer in number.

Our six failures were with adults and were of two types. Although they had received considerable amounts of leukocytes (five 10^{11} to 2×10^{11} per m², the sixth 7×10^{10}), the level of circulating leukocytes rose signifiicantly only twice—yet the temperature in these two patients failed to drop, and they soon died. Of the four others, three died quickly without the temperature being influenced; in the fourth, in whom the leukocyte count reached only 600 per mm³, the temperature fell and we later obtained a complete remission.

These failures appear to be associated with the seriousness of the patients' condition (adult myeloblastic leukemias, foci of infection, such as lung abscesses) rather than with the inadequacy of the quantities of leukocytes transfused or immunological rejection phenomena. It should also be observed that, though the influence of antigens of the red blood cell groups has been demonstrated for skin or kidney grafts, we have not seen any frankly worse results when we were compelled to transfuse ABO-incompatible leukocytes. These findings are in agreement with those of MORSE et al.

At least in our experience, secondary effects were negligible. We noted shiveriness and an increase of fever ten times, and during the transfusion itself a transient rash appeared three times, sometimes itchy.

We observed no marked hemolytic phenomena. So far we have seen no signs indicative of the secondary syndrome, (graft-versus-host reaction).

The value of leukocyte transfusions from chronic myeloid leukemia thus seems very great. They arrest very serious infections against which antibiotic treatment alone is powerless. Because of the presence of donor Philadelphia chromosomes in a great many mitoses in the marrow of three patients (in one in whom it was found in abundance three weeks later), it can be stated unequivocally that the donor's granulocytes were indeed responsible for the good results. The effectiveness of these granulocytes against micro-organism is an evidence that they keep their anti-infectious characteristics in chronic myeloid leukemia.

In sum, the variety of the problems raised by the restoration of the blood picture in patients with acute leukemia treated with rubidomycin or another drug or drugs causing aplasia is clearly to be seen.

From our experience of treating nearly 800 acute leukemias with rubidomycin alone or in combination with other drugs, it seems to us important to prepare for restorative measures before embarking on treatment: transfusions of platelets, isolation in a sterile room or bubble, summoning and hospitalization of patients with chronic myeloid leukemia and the same ABO and Rh blood groups as the patients with the acute leukemia. Another preventive measure is to eliminate, if possible, any foci of infection—particularly oral or dental—in patients due to undergo treatment that might cause aplasia.

When remission has been obtained, elimination of the foci of infection should be an absolute rule. Remissions may be very prolonged, and maintenance treatment favors serious and sometimes fatal infections, since the persistence of chronic infections may be the source of fresh septicemia.

Relapses will lead to the renewal of vigorous treatment capable of causing aplasia.

Chapter 7

Therapeutic Episodes and Accidents

The accidents caused by rubidomycin can be classified under three headings:

1. Marrow insufficiency, often encountered and often fulminant. Experimental study did not give an accurate picture of its frequency and abruptness of onset and course.

2. Cardiac accidents, indicated by the experimenters (page 90), occurring particularly as a result of prolonged treatment.

3. Others, rare and of little importance.

1. Marrow Insufficiency

Marrow insufficiency is seen especially in the treatment of acute leukemia. When rubidomycin is used in the treatment of sarcoma or cancer, a careful watch over the blood picture makes it possible in most cases to stop the treatment or reduce the amounts injected before severe marrow aplasia sets in. In the treatment of acute leukemia, on the other hand, aplasia sets in rapidly and is very severe. The risk must, however, be accepted, since remission only occurs after the phase of marrow insufficiency.

a) General Characteristics

The clinical manifestations of marrow insufficiency, with disappearance or marked fall in the number of nucleated elements in the marrow and extreme cytopenia of the blood, are very varied.

Aplasia is sometimes tolerated to a remarkable degree, though the marrow is denuded and only a few hundred leukocytes can be counted in the blood. After some days or a week or two at most, regeneration occurs without entailing any serious consequences, the patient having no temperature and not showing any hemorrhagic syndrome. It seemed to us that increase in the number of platelets is one of the first signs of marrow regeneration, followed by remission when the first course of rubidomycin has been adequate and the aplasia has not been too great or too lasting (see page 116, β).

In the patient with agranulocytosis, however, the clinical picture is more often dominated by infectious episodes. A sudden rise in temperature to 40° C, shiveriness, and sudden collapse are evidence of the onset of a septicemia most often caused by gram-negative organisms. A severe jaundice or renal failure may complicate the septicemia and carry the patient off in a few days. Or the temperature may remain high and irregular for a prolonged period and the organisms isolated be unequally sensitive to antibiotics; then the various forms of symptomatic treatment (page 79) must aim at limiting the effects of severe marrow insufficiency and combating the septicemia. Transfusions of fresh whole blood and platelet concentrates almost always prevent hemorrhage.

Perfusions of very heavy doses of antibiotics are in most cases inadequate, and only by transfusions of white cells from donors with chronic myeloid leukemia in a myelocytic phase can the white cell count be raised above 1000 per mm³ and the polymorphonuclear count above 800 per mm³. When the white cell counts begin to rise, the septicemia improves and the temperature falls rapidly (page 85). In general, early aplasia at the beginning of the attack treatment of the leukemia is less dangerous than aplasia occurring or becoming worse at a later stage, this being in most cases a poor prognosis (Table 31). In acute myeloblastic leukemia fatal aplasia is much

Table 31. *Time of occurrence of aplasia in acute myeloblastic leukemia treated with rubidomycin*

	Fatal aplasia	Non-fatal aplasia		Total
		Complete remissions	Incomplete remissions	
Early (first 3 weeks)	4	22	2	28
Late (4th—5th week)	10	5	2	17
Very late (6th—7th week)	7	8	4	19
Total	21	35	8	64

more common in patients who have had several courses of rubidomycin with total doses exceeding 10 mg/kg. We had four fatal aplasias in 28 patients and 22 complete remissions with a total dose of rubidomycin of less than 10 mg/kg; but when the total dose was above 10 mg/kg, we had 17 fatal aplasias and 13 complete remissions in 36 patients. The latter series included forms of the disease which are less responsive to rubidomycin but, as this cannot be known until treatment is begun, management of the drug is complicated.

Finally, the fatal issue may occur at a very late stage, say the fifth or sixth week. The aplasia has been overcome, the marrow blast cells have disappeared or greatly diminished in number, but the septicemia acquired during the period of aplasia drags on and ends tragically in death during the stage of regeneration. There are even some cases where remission occurs and the septicemia is cleared up, and then a secondary focus of infection causes the death of the patient within a few weeks or months of remission (Fig. 61, page 139).

b) Special Characteristics

Certain features give to the marrow insufficiency due to rubidomycin a special character of its own.

First there is its frequency: severe aplasia is seen in almost all cases; second, its abruptness: in a few days a well-populated marrow is stripped bare and the white cell count may fall from 300 000 per mm³ to 300 per mm³. This abruptness, which is an expression both of the quick action and of the intensity of effect of rubidomycin, is of very great importance. When it is decided to institute treatment of acute leukemia with rubidomycin the risk of such accidents is accepted, and by very frequent examination of the blood and bone marrow an effort is made to detect marrow insufficiency and correctly assess its course. The various symptomatic treatments that will limit its effects should be ready at hand for immediate application.

A third characteristic of the aplasia caused by rubidomycin is related to the platelets. The effect on the production of megakaryocytes and platelets is less severe, less rapid in onset, and less commonly found than in the other series. With most other antimitotic drugs, on the other hand, the changes in the platelets predominate. We have observed only four cases of fatal hemorrhage in nearly 800 cases of acute leukemia treated with rubidomycin.

A point that needs special stress is the maturation disturbances, which may be very considerable and occur at once or at the stage of regeneration. Maturation arrest may follow the stage of total aplasia and persist for several days with the marrow in the myeloblastic or promyelocytic phase.

The myelogram may then show the following successive stages:

1. total invasion by leukemic blast cells;
2. diminution in leukemic blastosis;
3. total aplasia;
4. reappearance of blast cells in the hypocellular marrow.

This is very important, for marrow blastosis at this stage does not necessarily mean an early relapse, the effect of inadequate treatment, but may be due to an overestimate of the blastosis in the still impoverished marrow (the proportion of blast cells counted in this case is certainly subject to a much greater range of error than when the marrow is rich in cells), or else maturation arrest due to the treatment and affecting the myeloblasts.

The ensuing myeloblastosis is generally moderate, affecting only the marrow (10—15%). It may be very marked, with 30—40% of nucleated elements in the marrow, and it may affect the blood, with 5—15% of myeloblasts present.

It is very often difficult to interpret the significance of these myeloblasts. If the leukemia under treatment is lymphoblastic, it is sometimes possible to make a diagnosis if the blast cells are still identifiable in spite of the treatment.

If the leukemia is myeloblastic, it may be very difficult to ascertain whether the condition is leukemic myeloblastosis or a maturation disturbance. The most refined cytological analyses do not always provide a solution, so that here again the clinician must pick a way between the difficulties, as he has to among the difficulties mentioned above (page 87). If in doubt, it is reasonable to postpone new injections of rubidomycin for two or three days.

The following example will give some idea of the importance of this regeneration blastosis:

DEV ... André, 39 years of age. This patient relapsed for the first time after 10 months of complete remission from acute myeloblastic leukemia, the remission having been obtained by a combination of prednisone, 6-mercaptopurine, methyl glyoxal, and cytosine arabinoside. The relapse was treated with rubidomycin, 13 mg/kg in nine days. From the 12th to the 21st day the marrow aplasia was complicated by a very severe septicemia and candidiasis; on the 20th day the marrow was denuded, and there were 400 white blood cells per mm³. On the 27th day, while the temperature had fallen after a transfusion of white blood cells, the marrow was still poor in cells, with 50% hemoblasts (Fig. 33); but no fresh injection of rubidomycin was given, because of the recent aplasia on the one hand and the absence of Auer bodies in the blast cells on the other. On the 32nd day the marrow was of normal richness and contained only 1% of blast cells but 24% of elements of the red series and 57% of granulocytes. The blood picture showed 11.5 g of hemoglobin, 230 000 platelets, 2200 white blood cells. Complete remission was obtained and continued for nine months.

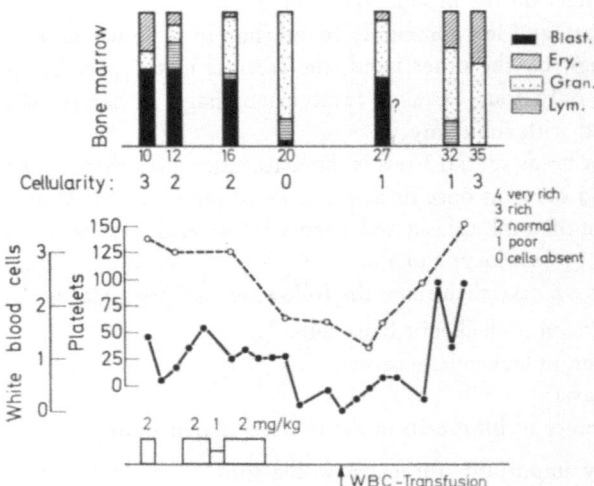

Fig. 33. First relapse in acute myeloblastic leukemia, treated with rubidomycin (13 mg/kg in nine days). Bone marrow poor in blast cells on 27th day. Remission on 32nd day (regeneration blast cells)

The following example shows the speed with which aplasia occurs after the injection of rubidomycin and vincristine and the need for a very careful watch over the blood picture and the state of the marrow during these very rapidly active treatments.

BOU ... Didier, 16 years of age. This patient had acute lymphoblastic leukemia of the hyperleukocytic form, with 380 000 white blood cells per mm³, of which 96% were lymphoblasts, 1 500 000 red blood cells, and 80 000 platelets. The sternal marrow had been invaded by elements that were 100% lymphoblastic. On 23 August 1966 the patient was placed on a treatment combining prednisone, vincristine, and rubidomycin (page 133). On 30 August the white cell count was 1400 per mm³ and the marrow was impoverished, containing 26% of hemoblasts. Because of the very rapid fall in the number of leukocytes and the possibility that the effect of the injection of rubidomycin and vincristine would be delayed up to the tenth day, the second injection that had been intended was not made. On 5 September the marrow was completely denuded and the white cell count was 200 per mm³. On 10 September the marrow was normal, the white cell count was 2,500 per mm³ and the proportion of polymorphonuclear cells was 54%. This complete remission was obtained on the 17th day of treatment by a single injection of rubidomycin and vincristine.

Thus, in the treatment of leukemia with rubidomycin, a difficult path must constantly be threaded between inadequate but not very dangerous doses and higher and more effective but often dangerous doses.

2. Cardiac Complications

Rubidomycin is capable of causing formidable cardiac complications [159, 214]. This disturbing characteristic had been recognized in the first experimental studies through its chronic toxicity to rabbits at three months on a dose of 1 mg/kg per day (whereas neither rats nor dogs showed any myocardial abnormality) [164, 165]. The frequency and gravity of these accidents should be neither overestimated nor underestimated. By correctly assessing the situation and judging the exact indications and doses for rubidomycin, the number of accidents it causes can be greatly reduced [26].

A study of the literature [36, 136, 186] enables the cardiac complications to be classified under two headings:

1. cardiac complications observed during prolonged treatment and undoubtedly due to rubidomycin; and

2. cardiac complications observed in other than prolonged treatment, the etiology of which deserves discussion.

a) Cardiac Complications Observed during and after Lengthy Treatment

These complications were mostly observed during the first clinical trials, when the toxicity of rubidomycin was poorly understood. They occur mostly in children with acute lymphoblastic leukemia or, less commonly, myeloblastic leukemia treated for several weeks with rubidomycin on total doses above 30 mg/kg. There is usually a lapse of four to six months between the institution of treatment and the appearance of the first signs and symptoms.

Clinically, the first disturbances are sinus tachycardia, which seems to be the alarm signal, and an enlarged liver, which rapidly comes to the forefront in the clinical picture. The signs of left ventricular insufficiency are less evident, and dyspnea is of variable intensity, there being no marked dyspnea when the patient is lying down. Crepitant rales may be audible at the base of the lungs on both sides. Attacks of acute pulmonary edema may occur, often being indicative of cardiac insufficiency. In three cases out of four a chest roentgenogram shows some degree of pulmonary stasis and a constant increase in the volume of the heart. The ECG confirms the sinus origin of the tachycardia; only once have we observed atrial tachycardia, and it appeared in a patient on digitalis and disappeared when it was stopped. The P waves may be of the pulmonary type, the P-R interval remains within normal limits, and the axis of the QRS complex seems—according to our observations—to be unusual (-150^c, -90°, -15° with S 1, S 2, and S 3 showing a retarded peak). These axis deviations recall those observed in acute cor pulmonale or in some infarcts. Repolarization disturbances have nothing specific: the waves are of low voltage in a rather diffuse way and there is a disputable excessive elevation of the ST segment in the right precordial lead. On the other hand, there is no obvious disturbance of conduction, and the QRS complex remains between 0.07 and 0.08. Biological investigations have not provided us with

Table 32. *Fatal cardiac accidents caused by rubidomycin*
(ALL: acute lymphoblastic leukemia; AML: acute myeloblastic leukemia)

Name of patient	Age	Diagnosis	Total dose	Start of treatment	Start of disturbances
BEN., J.	6 years	ALL	42 mg/kg	May 1966	13. 10. 1966
BIS., J.	10 years	AML	40 mg/kg	June 1966	30. 11. 1966
LEB., H.	6 years	ALL	32 mg/kg	July 1966	5. 12. 1966
VIG., L.	10 years	ALL	42 mg/kg	August 1966	December 1966
NEU., M.	8 years	ALL	45 mg/kg	June 1966	December 1966

any exact information. The development is rapid and progressive. Cardiac tonics seemed to us to have little effectiveness and, in particular, had no effect on the tachycardia.

We summarize below our first five observations (Table 32).

OBSERVATION No. 1 (Ben... J., 6 years of age). This little girl had been treated for acute lymphoblastic leukemia for 27 months. In January 1964 complete remission was obtained in four weeks by prednisone alone and maintained for 18 months by 6-mercaptopurine. In July 1965 the first relapse was treated with a combination of prednisone and vincristine and a complete remission was obtained by a combination of methotrexate and 6-mercaptopurine, interrupted by five reinductions with prednisone and vincristine. The second relapse occurred in June 1966, and remission was obtained in a month with 18 mg/kg of rubidomycin. The maintenance treatment was to be a weekly injection of 2 mg/for 12 weeks and the total dose in four months to be 45 mg/kg. On 13 October the child was admitted to hospital as an emergency with signs of cardiac insufficiency that had been appearing

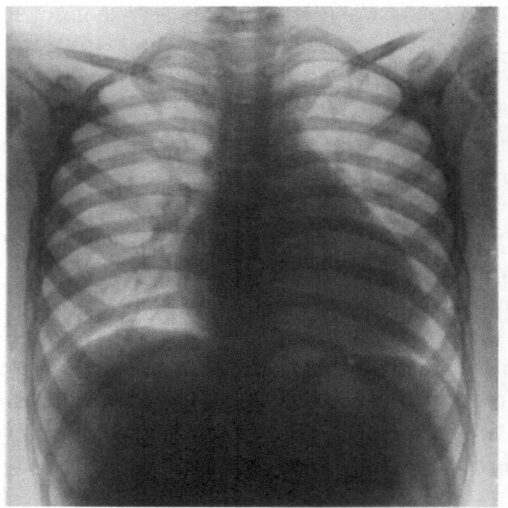

Fig. 34. Ben... J. Six years old. Overall increase in heart volume compared with previous films. Left inferior angle prominent

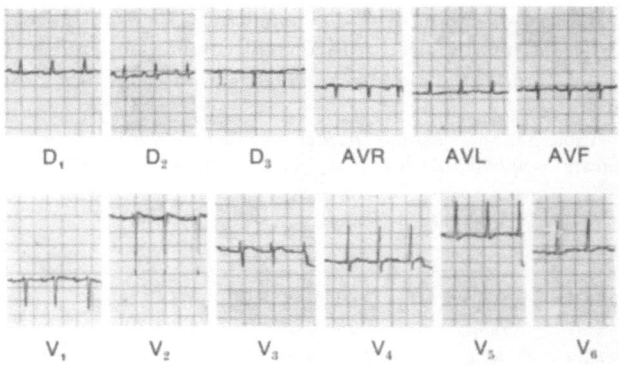

Fig. 35. Ben... J. ECG of 15.10.66 shows sinus rhythm of 150, P waves peaked in D 2, diphasic in Vl. QRS axis at —15°

progressively for a couple of weeks. On admission she had a heart rate of nearly 200, gallop rhythm, dyspnea with cyanosis, and an enlarged and painful liver with hepatojugular reflux. Her blood pressure was 110/90. The roentgenogram (Fig. 34) showed an overall increase in the heart volume and a prominent left lower border. The pleura at the lung base were free, the ECG (Fig. 35) of 15 October showed a slowing down of the heart rate to about 150 and sinus rhythm, the P waves in the R2 lead a little pointed and in the V1 lead diphasic, the P-R interval 16/100″, and the QRS axis at approximately −15°. There were no conduction disturbances. The T waves were flattened in all the leads (we had no previous ECG). Further investigations indicated a potassium deficiency with 3.3 mEq.

The course of the disease was marked by a transient improvement with digitalin and diuretics. On 20 October, however, the patient's pulse rate was 160, gallop rhythm returned, and the ECG (Fig. 36) showed an atrial systolic rate of 220 with some P waves blocked. When the digitalin was stopped, the extrasystoles were replaced by sinus tachycardia; the gallop rhythm and enlarged liver persisted. On 26 October the small patient returned home to her parents, and on 27 October she died.

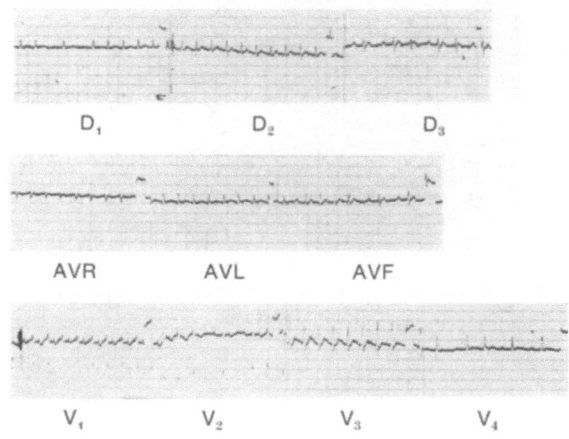

Fig. 36. Ben ... J. ECG of 20. 10. 66 shows atrial tachycardia with a rate of 220 and some P waves blocked

OBSERVATION No. 2 (Leb ... H., aged 6 years). This patient was a small boy who had been treated for acute lymphoblastic leukemia over a period of 34 months. A complete remission was obtained in June 1963 with prednisone and maintained for 20 months with 6-mercaptopurine. Then came the first relapse in April 1965. It was treated with prednisone and vincristine, then with 6-mercaptopurine and methotrexate. A third relapse in April 1966 was treated with vincristine, which led to serious neurological complications, and then with rubidomycin. After a dose of 9 mg/kg, a complete remission was noted on 8 July 1966. The maintenance treatment was to bring the total dose up to 32 mg/kg after five months of complete remission. On 19 October the heart was normal. A fourth relapse was noted on 30 November and the child was then treated with cytosine arabinoside. On 6 December he was found to have a heart rate of 140, gallop rhythm, and an enlarged and painful liver, and his blood pressure was 70/50. A chest roentgenogram showed an overall increase in the size of the chambers of the heart (Fig. 37) and pulmonary stasis. The ECG (Fig. 38) showed sinus rhythm of 130 and a P-R interval of 15/100″. The S1, S2, S3 appearance had a QRS complex axis that could not be calculated. In the V2 lead there were large S waves and in the V5 lead persisting S waves (8 mm). The T waves were flattened in the peripheral leads and in the left precordial leads. The potassium level was 3.6 mEq. On treatment with diuretics, digitalis, and opium derivatives, there was improvement, but it was transient and the patient died on 12 December 1966.

OBSERVATION No. 3 (Vig . . . P., aged 10 years). This boy had been treated for acute lymphoblastic leukemia over a period of 26 months. He was treated in February 1964 with prednisone, then with 6-mercaptopurine and methotrexate. After 16 months of

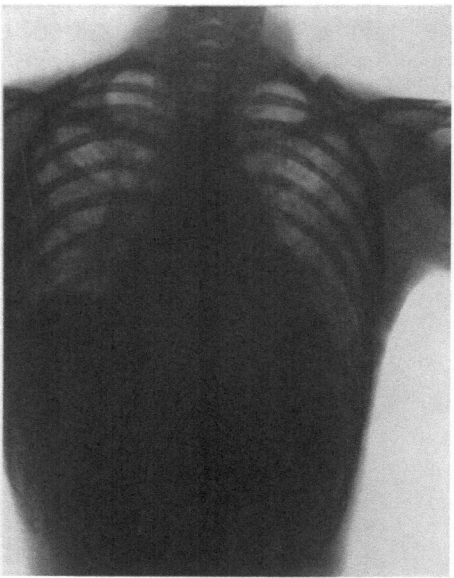

Fig. 37. Leb . . . H. Overall increase in heart chambers as compared with previous films. Pulmonary stasis

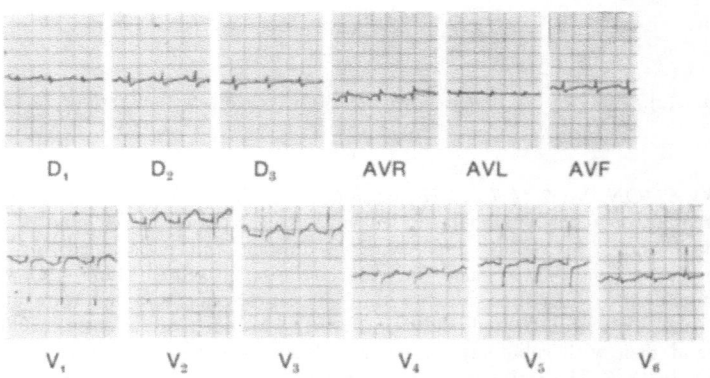

Fig. 38. Leb . . . H. Six years of age. Sinus rhythm of 130, PR 15/100, appearance S1—S2—S3

remission he had a relapse in August 1965, which was treated with prednisone and vincristine, then with 6-mercaptopurine and methotrexate separately, alternating with reinductions with prednisone and vincristine. A second relapse in February 1966 was resistant to three weeks of treatment with cytosine arabinoside, and he was then treated with rubidomycin (Fig. 48 p. 103), a complete remission being obtained on 16 April 1966 after a total dose of 6 mg/kg, administered in six days. The maintenance treatment brought the total dose of rubidomycin up to 30 mg/kg. On 17 October 1966, after seven months of remission, the patient presented with an enlarged and painful liver, which was the main feature of the clinical picture, accompanied by vomiting, a regular fast heart rate of 150, and a proto-diastolic gallop rhythm. Hepatojugular reflux was present, and the systolic blood pressure was 80. The situation was improved by the injection of a half ampoule of lanatoside C. The

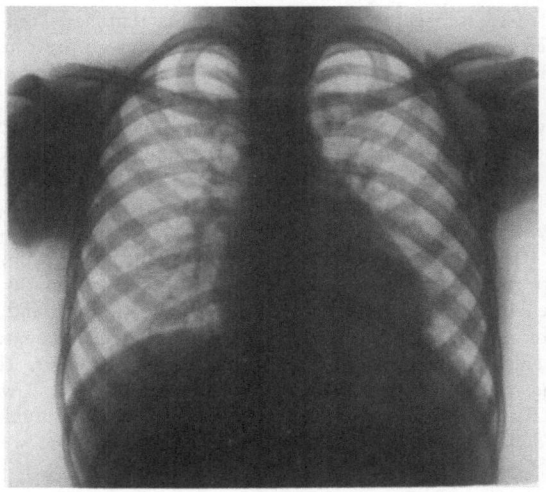

Fig. 39. Vig... P. Ten years of age. Overlap of right inferior angle without double contour. Bilateral pulmonary congestion

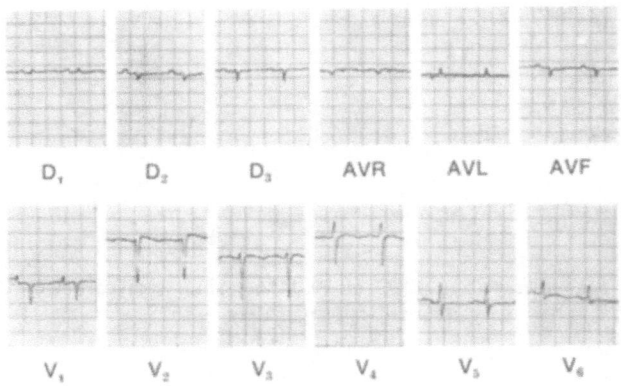

Fig. 40 Vig... P. Sinus tachycardia of 150. Axis at —90°. P2 < P3 < P1. P diphasic and pointed in V1

roentgenogram (Fig. 39) showed an enlarged heart with one edge of the right inferior angle without a visible double contour and a prominent left inferior angle dipping behind the diaphragm, the middle arch straight, and considerable pulmonary congestion on both sides. The ECG (Fig. 40) showed sinus tachycardia with a rate of 150, a peaked P wave less than 2 mm but higher than P3 and P1, a normal P-R interval, a QRS axis of —90°, and diffusely flattened T waves.

The course of the disease was marked by a transient improvement under digitalin and diuretics, but the tachycardia persisted.

A relapse occurred on 25 October, when rubidomycin had been stopped for ten days. It was again treated with cytosine arabinoside. Cardiac tonics were poorly tolerated, but opium derivatives improved the enlarged and painful liver. On 3 December the liver again gave trouble, and lanatoside C had to be given. On 4 December the blood pressure dropped to below 80 mm Hg. A trial with Insuprel improved the cardiovascular condition but increased the heart rate to 130, which made the administration of reserpine and digitalin necessary. The patient died on 5 December.

OBSERVATION No. 4 (Bis ... J., aged 10 years). The fourth patient was also a boy of ten years of age, treated for acute myeloblastic leukemia that had been going on for nine months and had been diagnosed in September 1965. There was no previous history of cardiac disturbances and the clinical and radiological examination of the heart at that time showed no abnormality.

Remission was obtained completely, after cytosine arabinoside used alone for three weeks had failed, by a combination of prednisone, 6-mercaptopurine, methyl-GAG, and cytosine arabinoside. The first relapse occurred in June 1966 and was treated with rubidomycin, five injections ensuring a complete remission. After a very short period of aplasia, lasting four days, the patient was put on maintenance treatment, which was to bring the total dose of rubidomycin up to 40 mg/kg. On 26 November he had slight dyspnea, and on 30 November a regular fast heart rate of 130 was observed, along with a gallop rhythm with a proto-diastolic peak. An enlarged and painful liver and cyanosis of the lips and extremities rapidly developed, but there was no sign of pulmonary edema. The roentgenogram on 3 December showed an overall enlargement of the heart affecting all the chambers. The ECG of 30 November (Fig. 41) showed sinus tachycardia with a heart rate of 130, P waves a little raised in D2, a normal P-R interval, no bundle branch block but a very unusual QRS complex axis in the neighborhood of —150°. In the precordial leads the R waves were small up to the

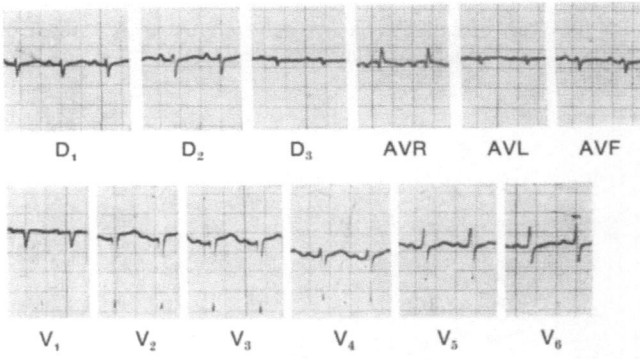

Fig. 41. Bis ... J. Ten years of age. ECG of 30. 11. 66. Sinus tachycardia of 130. QRS axis very unusual, approximately —150°

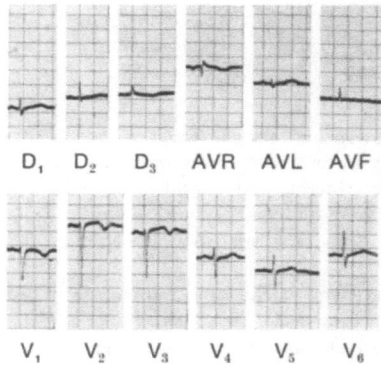

Fig. 42. Bis ... J. ECG of 19. 10. 66. QRS axis at +60°

V4 lead and the S wave deep up to V5 and V6, in which R=S. The T waves were diffusely flattened. It is interesting to compare this tracing with an ECG recorded on 19 October (Fig. 42), which was already abnormal, in the precordial leads in particular. The T waves were low, the QT interval was lengthened, but the QRS axis was about +60°, so that in a month it changed from +60° to −150°. The other investigations were not very informative: the potassium level was 4.6 mEq, the serum transaminases SGOT and SGPT were normal, and the serum lactic dehydrogenase was a little up (250, normal less than 200). After a transient improvement in the hospital cardiovascular department on cardiac tonics, diuretics, a salt-free regimen, and oxygen—which reduced the cyanosis but had no effect on the enlarged liver or the heart rate (between 100 and 120)—rapid deterioration set in, as was to be expected, and the patient died on 6 December in a cyanotic condition and a state of collapse, in spite of the administration of vasopressor agents and hydrocortisone hemi-succinate.

OBSERVATION No. 5 (Neu ... M). This patient suffering from acute lymphoblastic leukemia resistant to prednisone, vincristine, methotrexate, and 6-mercaptopurine had a fourth relapse. In June, July, and August 1966 three courses of rubidomycin produced only three incomplete remissions of short duration. The total dose of rubidomycin administered amounted to 45 mg/kg.

Later cytosine arabinoside was tried without effect. The patient was then lost sight of. He died at the end of December 1966 from cardiac insufficiency with dyspnea, an enlarged liver, and a heart rate of 160—a picture that, according to the written report of the doctor in charge, resembled that observed on the four occasions he was treated by us.

Two remarks are worth making:

1. These complications were not constant even in subjects receiving doses higher than 30 mg/kg. At a time when these cardiac complications were poorly understood, 11 children received doses higher than 30 mg/kg, and fatal cardiac complications occurred in only five. Of the six others, four died from a relapse of their disease without showing any heart abnormalities in periods ranging from two to six months after the institution of treatment; but since there was no histological test, this cannot be affirmed categorically. The other two are still in remission at the time of writing. We have observed no change in a clinical, radiological, and electrical examination of the heart, at periods respectively nine months and six months after the institution of treatment.

Thus patients subjected to high and prolonged doses of rubidomycin fall into two classes: some succumb to fatal cardiac complications, while the others have so far shown no signs of heart trouble, although it may well be that the period of observation is still too short.

Anatomical study. A post-mortem examination was made in one case [186].

OBSERVATION (Vig ... Patrice, aged 10 years). In this case the post-mortem examination showed diffuse generalized purpuric lesions related to marrow aplasia associated with a moderate infiltration of blast cells. There were some septicemic areas in the intestine, and the liver showed a diffuse centrilobular degeneration. Well-marked lymphoid hypoplasia was seen in the spleen.

The heart demonstrated multiple lesions: (Figs. 43, 44, 47, see p. 145)

1. a small serohemorrhagic pericardial effusion with epicardial purpura, representing the local manifestation of the hemorrhagic lesions observed elsewhere;

2. a right and left ventricular dilatation that reduced the wall on the left side to a thickness of 1/2 cm (Fig. 43) although, given the age of the patient, the weight of the heart (175 g) was approximately normal; and

3. multiple thromboses adhering to the walls of both ventricles.

It will be observed that there was no valvular thrombosis or lesion whatever, no obliteration of the pulmonary artery, and no deformation of the heart or great vessels.

Histological examination confirmed that the thrombosis was intraventricular (Fig. 44) and had a more or less completely organized base. There was in addition a more or less substantial fibrous thickening of the endocardium, clearly derived from the connective-tissue

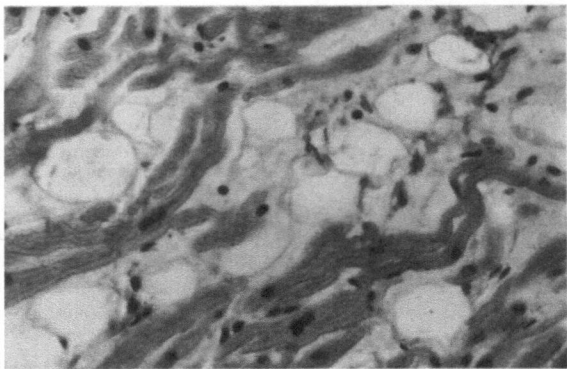

Fig. 45. Fibrous tissue lining the interior of the heart chambers, formed by the organization of the mural thrombus (magnification ×25)

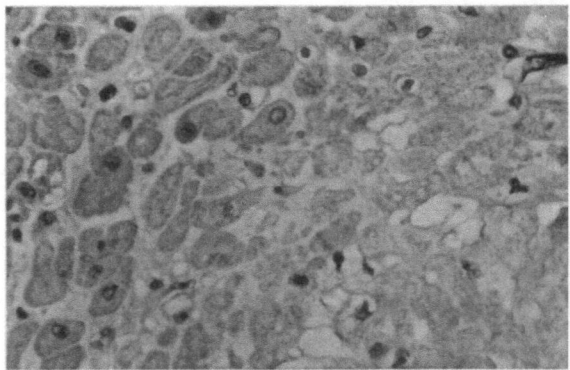

Fig. 46. Edema and interstitial fibrosis of the myocardium (magnification ×250)

organization of the parietal thrombi (Fig. 45). Here and there were to be found subendocardial fibrinous deposits. The myocardium was the site of an interstitial myocarditis consisting either of edema and congestion or of sclerosis (Fig. 46); in places, especially on the internal aspect of the myocardium, there were lesions affecting the myocardial fiber itself— granular degeneration, vacuolar degeneration, and islets of cytolysis (Fig. 47). The blood vessels, nerves, and conduction system were histologically normal.

To summarize: this patient had on the one hand hemorrhagic and septicemic lesions related to his marrow aplasia, on the other, complex cardiac lesions. The ventricular parietal thromboses were unusual, but can be explained by stasis, itself due to the ventricular dilatation, which seems to be connected with the myocarditis.

Attempt at a physiopathological interpretation. The mechanism of these cardiac complications has not as yet been elucidated. A hypothesis put forward is that they are an indirect effect through the autonomic nervous system, causing a fall in the peripheral resistance and an increase in the heart output; but this hypothesis has not been confirmed. Nor have pathological examinations confirmed the hypothesis—put forward because of the predominance of signs implicating the right heart—that the complications are due to thromboembolism. Experimental findings suggest it is much more probable that rubidomycin acts directly on the myocardium.

b) Cardiac Complications Observed during Less Prolonged Treatment

The observations collected, far from having the unity of those of the preceding group, are very diverse in their nature (Table 33).

Table 33. *Cardiac accidents with rubidomycin in terms of the total dose employed*

Dose received per kg body weight	Number of patients	Undoubted accidents	Doubtful accidents
Over 30 mg	13	5 (fatal)	5 (2 fatal)
Between 10 and 30 mg	135		
Less than 10 mg	266		4 (3 fatal)
TOTAL	414		

Accidents observed in older patients or in patients with a history of previous heart disorders (coronary disorders, narrowing of the aorta by calcification, hypertension, left ventricular insufficiency). In these cases either myocardial infarcts or senile myocarditis were the complications observed, and it was difficult to decide whether rubidomycin was implicated or whether it played a contributory role. We have collected five observations on such cases. In the two where autopsy was possible, it confirmed that the lesions affected the coronary vessels or were nonspecific and characteristic of the senile heart. In the same period we saw approximately the same number of cardiac complications of this kind among patients in our department with other conditions such as chronic lymphoid leukemia, a disease common in the old and not treated with rubidomycin.

Complications occurring during leukemia complicated by severe septicemia. Whatever the antileukemic treatment employed, it is not exceptional for septicemia occurring at the phase of aplasia to be suddenly complicated by cardiovascular collapse. These complications have also been observed with leukemia patients treated with rubidomycin, and it is very difficult to determine whether rubidomycin increases the frequency and adversely affects the course of the cardiac condition. We have seen three cases of this kind.

Isolated cardiac complications. As well as episodes in the aged, in individuals with a previous history of heart trouble, and complicating septicemia in the period of aplasia, there are isolated cardiac episodes that may occur in patients who have not received high doses of rubidomycin. They seem to appear most often in the form of acute or subacute pulmonary edema, which may or may not be followed by considerable cardiac insufficiency. They always do well. It is very probable that rubidomycin is to some extent responsible, but it is not without interest to note that the symptoms are not quite the same as those of the episodes where high doses of rubidomycin are involved. These occurrences have been observed by the various teams of workers who have employed rubidomycin. In our personal series, we have noted two cases only out of more than 800 acute leukemias treated with rubidomycin, often in very moderate doses.

c) Conclusion: Treatment and Attempted Prevention of Cardiac Complications

As we have seen, the classical treatment of cardiac insufficiency is not very effective. Nevertheless, digitalin, diuretics, oxygen—if there is any considerable cyanosis—vascular stimulants (in cases of collapse), and bleeding or quick-acting diuretics (in cases of acute pulmonary edema) are indicated.

In spite of the uncertainties we have recorded above, there are generally effective preventive measures that can be taken. They fall into three classes related to 1. the choice of the patients treated, 2. surveillance, and 3. limitation of dose.

Choice of patients treated. In principle, patients with a previous history of cardiac disturbances and those above the age of 75 years should not be treated with rubidomycin. A clinical, radiological, and electrocardiographic examination of the heart should always be made before treatment is begun and repeated at regular intervals when treatment requires cumulative doses greater than 15 mg/kg [159].

Surveillance. A very strict clinical, radiological, and electrocardiographic surveillance should be instituted. Here the close collaboration of the cardiologist and the hematologist is essential. The surveillance will include: (a) a careful watch on the heart rate, any tachycardia being considered as an alarm signal; (b) a careful watch over the heart volume by repeated chest X-ray; and (c) regular examination by ECG, with a particular look-out for the appearance of any change, especially disturbances of repolarization.

Limitation of dose. It is very important not to exceed a total dose of 30 mg/kg. It is prudent not to go beyond a total dose of 25 mg/kg except in cases of urgent need.

Rubidomycin can therefore never be employed for maintenance treatment. It can only be used as attacking treatment and in reinductions for limited periods.

3. Other Complications

a) Local Complications

Local complications may be extremely serious if the injection of the drug is not strictly intravenous. We have seen in two cases inflammatory lesions and extensive edema from the bend of the elbow to the lower third of the forearm. These lesions became necrotic and sloughed, and after a course of three weeks fibrous retraction occurred as a sequela and in one case several skin grafts were necessary. Three other patients had local lesions but they healed up uneventfully; they were inflammatory and very painful but much less extensive.

In spite of taking the precaution of injecting the rubidomycin into the tubing during a transfusion of physiological solution, accidents are possible. They mostly occur in patients who have had leukemia for a long time with several relapses and have had to undergo many intravenous catheterizations.

b) Miscellaneous Complications and Accidents

1. Alopecia, often complete, is very common towards the end of attack treatment. We have observed various degrees of it in all our patients treated for acute leukemia.

It disappears fairly rapidly during a remission, but may reappear very fleetingly at reinduction involving the combination of vincristine and rubidomycin.

2. Very painful buccal ulceration with stomatitis has been observed with severe and long-lasting aplasia. It healed quickly.

3. On three occasions convulsive attacks occurred during periods of very severe aplasia complicated by septicemia.

4. In general, our routine examination of hepatic and renal function revealed no changes either towards the end of attack treatment or at reinduction, or on the rare occasions when maintenance treatment was given.

In one patient hepatitis occurring after transfusion cleared up in the usual time, at the third month of maintenance treatment with rubidomycin (which was continued for another two months).

The renal complications that from experimental study seemed to constitute a danger appear to be very rare. In two patients, however, we observed signs of tubular change. In one case a complex tubular syndrome developed, akin to adult acquired Fanconi's syndrome. The first clinical signs, associated with hypocalcemia (7,5 mg calcium per 100 ml), appeared following septicemia treated with a combination of penicillin and colistin. This patient had a complete remission of his acute myeloblastic leukemia that continued for ten months, the renal disturbances regressing almost entirely. In another case the patient had an acute tubular nephritis (confirmed by renal biopsy) with anuria; this condition developed during a period of aplasia after treatment with the same combination of antibiotics. The renal lesions regressed, but the acute myeloblastic leukemia recurred in a rapidly developing form, causing the death of the patient after diuresis had returned.

In four cases convulsive attacks were noted during periods of very severe aplasia complicated by septicemia. They occurred shortly prior to the death of the patients.

Chapter 8

Treatment of Acute Lymphoblastic Leukemia with Rubidomycin Only

Acute leukemias were first treated with rubidomycin alone, which enabled the effects of the new drug to be assessed as accurately as possible. Then they were treated with combinations of drugs, rubidomycin being used in association with others.

1. Preliminary Remarks

a) Techniques Employed in Hematologic Diagnosis

All the diagnoses were made in our laboratory on bone marrow and blood smears stained with May-Grünwald-Giemsa stain. The cytological differentiation of the acute leukemias into acute myeloblastic, promyelocytic, lymphoblastic, or undifferentiated was made on the one hand on the basis of the usually accepted cytologic criteria after panoptic staining [24], on the other after the application of cytochemical techniques (PAS staining of the peroxidases, Sudan black staining of the esterases) in accordance with the criteria defined by HAYHOE [116].

b) Definition of Complete Remission

The criteria defining complete remission are as follows, and conform to the international conventions and to the recommendations of the Clinical Studies Panel of the Cancer Chemotherapy National Service Center [124]:

1. Clinical state — especially normal spleen and glands
2. Blood:
     hemoglobin > 14.5 g
     number of platelets > 180 000/mm³
     number of leukocytes > 4000/mm³
     number of polymorphonuclear leukocytes > 2000/mm³
     no blast cells
3. Bone marrow:
     normal cell population
     number of blast cells less than 6⁰/o
     number of lymphocytes less than 20⁰/o

These criteria should now be both stricter and less strict. Stricter because the integrity of the cerebrospinal fluid is necessary and should be checked, less strict because the blood criteria are too rigid.

In principle, the above criteria ought only to be applied to a first remission. They cannot be required later because the purpose and consequences of many maintenance treatments are in fact to keep the blood in a state of moderate cytopenia. The very many forms of drug treatment employed for these patients at successive relapses explain the relative marrow insufficiency that persists after the attack treatment and the very slight indication for the institution of maintenance treatment, which should take effect before complete marrow regeneration because of the risk of an early fulminating relapse. The necessary relaxation in the criteria affects the blood only. The marrow blast cell level should remain less than 6⁰/o. We have considered patients in complete remission when their marrow was normal (M 1), their clinical examination absolutely normal (P 1), and they were without any symptom of the disease and had fully recovered their physical strength (S 1). On the other hand, their blood picture could remain at the standard stage 2 (H 2), with the hemoglobin ranging between 11 and 14.5 g, the platelet count between 150 000 and 180 000, the white cell count between 3000 and 4000, and more than 1500 polynuclears per mm³.

According to these criteria a remission can be considered as complete with a blood picture of M1 H2 P1 S1. We consider this to be extremely important because rubidomycin causes marked marrow erythroblastosis, with abnormal erythroblasts (page 147). We had already made the same observations when we treated acute leukemia with cytosine arabinoside.

c) Preliminary Observations

In our initial ignorance of effective and best-tolerated doses and of the toxicity of rubidomycin to the blood and in general, we employed it only as a last resort on patients who were not only resistant to known drugs—prednisone, vincristine, methotrexate, 6-mercaptopurine and, in certain cases, cytosine arabinoside and cyclo-phosphamide—but were also in a very poor general condition [131]. In these first cases we observed that the peripheral hemoblasts rapidly disappeared and marrow blastosis declined, and this encouraged us to pursue the study of this drug in forms of acute leukemia that were less immediately critical [26, 132].

2. Dosage

There is general agreement that in these cases the treatment should be short and fairly intense.

The injection dose should be 1 mg/kg.

The injections should be daily and the total dose given should be 3—20 mg/kg. In attack treatment there should be one or two courses, and the maximum dose used should be 12 mg/kg; in a single treatment course the median dose should be 10 mg/kg, given over 3—6 days consecutively.

3. Study of a First Personal Series

Patients treated. Because of the high percentage of complete remissions obtained by the known treatments such as vincristine and prednisone, and the certainty of prolonging the duration of remission to a significant extent by the use of the usual antimetabolites, it was not possible at the beginning to employ the new drug in the treatment of fresh cases of acute lymphoblastic leukemia. Thus the first lymphoblastic leukemias treated with rubidomycin were serious forms that were refractory to other treatments or in relapse. The first series of cases that we treated, as described below, provides a good example of this particularly unfavorable selection [136].

The number of cases in which a sufficiently lengthy period of observation was possible was 38.

Of these, 21 were at their second, third, or fourth relapse and had become resistant to prednisone, vincristine, methotrexate, 6-mercaptopurine, and sometimes also cytosine arabinoside (Fig. 48).

Of the others, 10 were being treated in their first relapse. In the case of four of these, the combination of prednisone and vincristine had proved to be ineffective over a period of four weeks or more. In the case of the other six, the relapse had

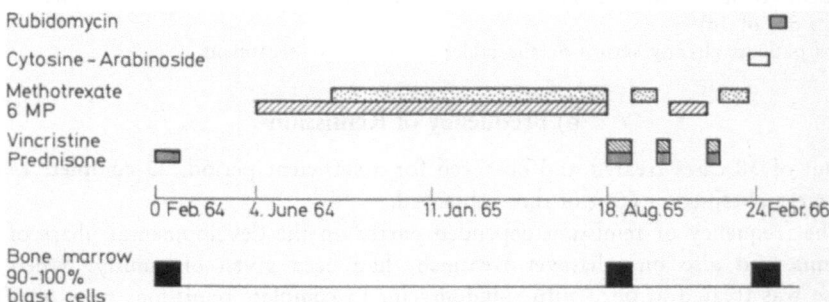

Fig. 48. Acute lymphoblastic leukemia treated for two years; second relapse. Complete remission obtained with rubidomycin after failure of cytosine arabinoside

occurred after a remission; maintenance treatment, as well as the combination of intermittent methotrexate and 6-mercaptopurine, had consisted of multiple reinductions with prednisone and vincristine combined, and we had therefore concluded that the patients were probably resistant to those drugs.

The remaining 7 were being treated on the first manifestation of the disease. Four had been treated ineffectively for a long time with prednisone and vincristine; one with prednisone alone in insufficient dosage for more than two months; and the last two, who were treated at once with rubidomycin, had presented difficult cytological classification problems and had initially not been considered as having lymphoblastic leukemia.

Thus two categories of patient can be distinguished:

1. those whose relapse was treated at once with rubidomycin (in all 20);

2. those to whom rubidomycin was administered after the failure of other treatment (in all 18).

a) Results

The results obtained are shown in Table 34.

Table 34. *Treatment of acute lymphoblastic leukemia with rubidomycin. Results obtained in attack treatment in terms of the stage of the disease*
(23 complete remissions out of 38, i. e., 60%)

	1st attack		1st relapse		2nd relapse		3rd relapse		4th relapse		5th relapse		Total
	a	b	a	b	a	b	a	b	a	b	a	b	
Complete remissions	2	1	5	2	4	3	3	2			1		23
Incomplete remissions		1	1^+				$1+1^+$		$1+1^+$				$3+3^+$
aplasia without blast cells				2	1^+	1		2					$5+1^+$
aplasia with blast cells		3			1		1	1					6
Resistant			1										1
	2	5	$6+1^+4$		$5+1^+4$		$5+1^+5$		$1+1^+$		1		$38+4^+$
	7		$10+1^+$		$9+1^+$		$10+1^+$		$1+1^+$		1		$38+4^+$

a: patients treated from the outset with rubidomycin (15 complete remissions out of 20)
b: patients treated with rubidomycin after failure with other treatments (8 complete remissions out of 18)
   +: patients already shown on the table; 2nd course of treatment

b) Frequency of Remission

Out of 38 cases treated and observed for a sufficient period, 23 complete remissions were obtained or 60% of those observed.

The frequency of remission depended partly on the developmental phase of the leukemia and also on whatever treatment had been given previously. When the relapse was treated at once with rubidomycin, 15 complete remissions were obtained out of 20 patients, i. e. 75%. When rubidomycin was given only after the failure of other treatment, the proportion of remissions was less than 50% (eight remissions in 18 patients treated). In such cases, as for the combination of predniscne and vincristine, the patients who had already been treated for a long time were much more delicate, recovery from aplasia was slower, bacterial infections were commoner and more dangerous, and remission was more difficult to obtain.

c) Complete Remission. Chronology, Quantitative Changes

The three facts worthy of note are:

1. the speed with which the blast cells circulating in the blood disappeared;

2. the extent and gravity of marrow aplasia; and

3. the frequent difficulty of adjusting the treatment at the end of the aplastic stage, owing to the often tricky problem of interpreting residual blastosis in an impoverished marrow.

The blast cells circulating in the blood disappeared quickly, even before the end of the period of attack treatment (Fig. 49).

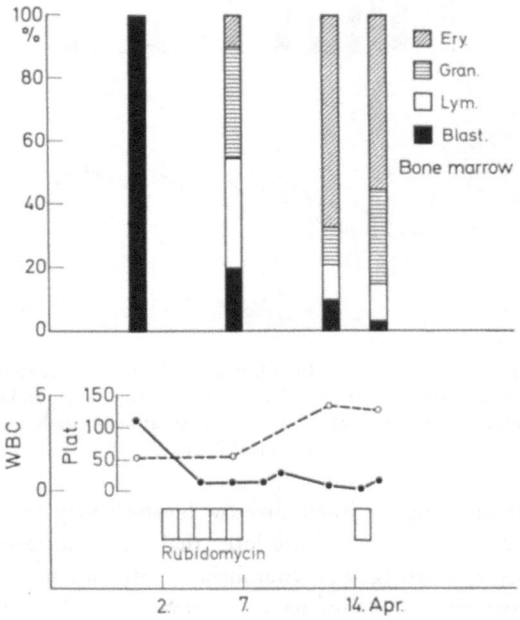

Fig. 49. Acute lymphoblastic leukemia, second relapse. Complete remission obtained on 14th day. Rapid disappearance of blast cells from blood and marrow (1 mg/kg/d)

At the end of the attack treatment, the marrow blast cells were fairly characteristic. The appearance of giant lymphoblasts occurred a little prior to aplasia and was an indication for stopping rubidomycin. Because of this careful cytological study, it was possible to adapt the dosage of the drug to each individual case (page 44).

Marrow aplasia set in between the 10th and 20th day and lasted on an average for 15 days. Peripheral leukopenia grew worse and was very severe, figures of from 400 to 200 white cells per mm³ being frequently observed. During this period symptomatic treatment—perfusions of antibiotics in high dosage, transfusions of fresh blood and platelets—was of major importance. When leukopenia persisted for more than 8—10 days the temperature rose, shivering fits set in, and septicemia was often observed, almost always caused by gram-negative organisms. At this stage of the development of the disease, transfusions of leukocytes from donors with chronic myeloid leukemia in the myelocytic phase were of great value.

The period at which adjustment of the treatment to suit the patient is difficult is when leukopenia starts to disappear, around the 10th—20th day, when attack treatment is coming to an end (Fig. 50).

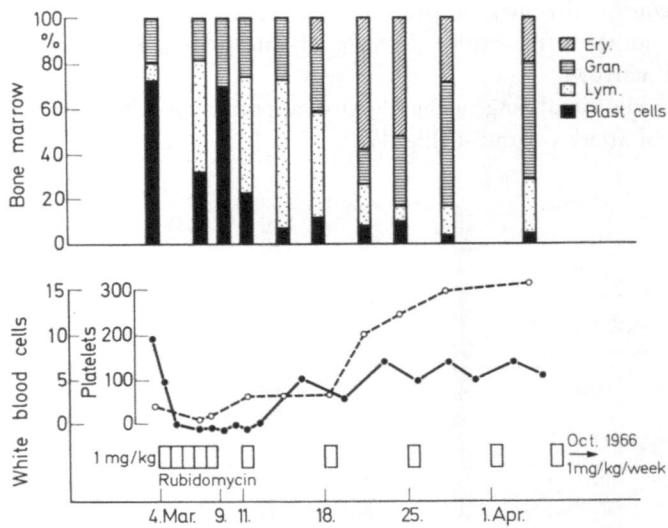

Fig. 50. Complete remission in acute lymphoblastic leukemia treated with rubidomycin. Course of blood picture during maintenance treatment for seven months on rubidomycin, 1 mg/kg weekly. Relapse ten days after arrest of treatment with rubidomycin owing to cardiac insufficiency

The marrow remains impoverished, marrow blastosis may affect between 5% and 40% of the cells, and it is only a little later that, in some cases, the marrow will spontaneously recover a satisfactory equilibrium with not more than 5% of blast cells. Thus the institution of maintenance treatment should not be too early and aggravate the aplasia, nor too late and jeopardize the remission. This period occurs between the 10th and 15th day of treatment, and treatment should start only after two successive examinations show that the white blood cell count is clearly rising and has reached or passed 1200 per mm³ and that the marrow is averagely rich in cells.

During this waiting period, we were compelled in the case of eight patients in whom the proportion of blast cells rose moderately, from 15% to 35%, in a cell-rich marrow to revert to attack treatment of short duration, administering 1 mg/kg daily for 2—4 days or giving one injection of 2 mg/kg before obtaining a complete remission after a fresh phase of aplasia with fever.

d) Incomplete Remissions

Among the incomplete remissions (three cases) were two in which a complete reduction of marrow blastosis was only transient—in one of them on three occasions. The third case was one of acute leukemia with cells that were difficult to classify, and it was resistant for five weeks to the combination of prednisone and vincristine. Each time rubidomycin was used in treatment a remarkable complete regression of considerable splenomegaly and very marked cervical adenopathy was obtained, while marrow blastosis persisted with blast cells ranging between 8% and 15% of the total.

e) Partial Failures (11 Cases)

α) Aplasia without Blast Cells

In five cases, and in a sixth at the second course of treatment (the patient had had a complete remission lasting five weeks, maintenance treatment with rubidomycin, and then a relapse), the treatment was active: the marrow blast cells disappeared, the marrow became deprived of cells, and death ensued with the clinical picture one of septicemia. These were cases treated at the outset of the study, probably with doses that were too high—2 mg/kg daily for five days—and with a maintenance treatment that was too early, when marrow regeneration had scarcely begun. It is reasonable to believe that if these patients had received a lower dose administered at a later stage, if more attention had been paid to the increased proportion of blast cells in the impoverished marrow, if the treatment with rubidomycin had been shorter, and if more vigorous recovery treatment had been given during the stage of aplasia, fatal episodes of infection would have been avoided and the number of complete remissions would have been greater.

β) Aplasia with Persistence of Blast Cells

In six other cases the action of the rubidomycin was too rapid, aplasia could not be controlled, and death was due to septicemia at a time when blast cells still persisted in the impoverished marrow. Four of these patients were being treated with rubidomycin after other treatment had failed; in three cases they were suffering from acute leukemia, refractory from the outset to treatment with prednisone and vincristine over more than six weeks.

f) Total Failure

Only in one case was primary resistance to rubidomycin observed. With first a five-day course of 1 mg/kg every other day, second a course at 2 mg/kg, and third a course at 3 mg/kg every third day with three injections, the peripheral blastosis disappeared, but only transiently, and there was no marked change in the marrow blastosis.

g) Duration of Remissions

The duration of the remissions is shown in Table 35.

Some remissions were very short, others longer. Three patients survived for 7, 4.5, and 3.5 months respectively on maintenance treatment with rubidomycin at a dose of 1 mg/kg weekly without hematologic relapse, then died from heart complications.

Rubidomycin was later not used alone as maintenance treatment but only for reinduction. Two patients in remission for ten and for four months respectively received more than 30 mg/kg of rubidomycin without mishap in that time.

Because of the initial seriousness of the cases treated and the great variety of forms the disease took in its development, it was impossible to assess the duration of the complete remissions obtained. This study will, however, assume much greater importance when the results are analysed of combined treatment applied at an early stage to acute lymphoblastic leukemias in their initial phases.

Table 35. *Duration of complete remissions (months) in acute lymphoblastic leukemia (23 cases) treated with rubidomycin by weekly injection (30 mg/m², i. e., 1 mg/kg)*

1st attack	30°+, 8°, 5°
1st relapse	20°, 4, 9, 3, 4°, 1, 1
2nd relapse	7 (IC), 3.5 (IC), 4, 2.5, 2, 2, (0.5), (0.5)
3rd relapse	4.5 (IC), 2, 1, 0.5
5th relapse	1

°   Patients whose maintenance treatment had been changed (Protocol 06 LA 66) after a cumulative dose of 600 mg/m² (20 mg/kg) of rubidomycin
IC Fatal cardiac insufficiency

h) Complications Caused by Rubidomycin

The accidents and complications caused by rubidomycin are described on page 87.

4. Two Examples of Rubidomycin Used Alone

To illustrate our personal experience of rubidomycin used as the sole form of treatment in acute lymphoblastic leukemia, we give below a summary of two particularly revealing cases. Three other cases are reported in connexion with cardiac mishaps (page 92).

CHA ... Eric, aged 3.5 years, who was in his fifth relapse in an acute lymphoblastic leukemia that had been going on for 22 months. The course of the case and the treatment the patient had been receiving are summarized in Table 36. When he had a relapse in March 1966, he had 56 400 white blood cells in his peripheral blood, of which 7% were polymorphonuclears and 80% hemoblasts, and the number of platelets was 370 000. His bone marrow was rich in cells, all hemoblasts, in sheets. He was given injections of rubidomycin from 14 to 19 March, in a dose of 1 mg/kg daily for the six days. On 29 March, ten days after the last injection, the blood picture showed 800 white cells in the peripheral blood, 11.5 g of hemoglobin, 10% of polymorphonuclears, 85% of lymphocytes, 5% of monocytes, and 20 000 platelets, while the very impoverished marrow contained only 10% of hemoblasts and 76% of lymphocytes.

Five days later the peripheral blood was normal—12.5 g hemoglobin, 3000 white cells, 50% polymorphonuclears; and the marrow was normal—5% hemoblasts, 64% of blood elements of the granulocyte series, 22% of the erythrocyte series, and 9% of lymphocytes. Maintenance treatment with rubidomycin was pursued for six weeks (in a dose of 1 mg/kg weekly). A relapse was treated with higher doses of rubidomycin (2 mg/kg daily for five

Table 36. Acute lymphoblastic leukemia. Duration of successive remissions during the course of 2 years in terms of treatment employed

Name: CHAP
First name: Eric
File no.: 61/1732

Acute leukemia course

Age: 3½ years
Sex: male
Date of complete remission: June 1964
Duration of life: 2 years

Protocol no.	Relapses	Cl	He	Me	Tu	Treatment Induction	Maintenance	Reinduction	Duration of complete remission
01 LA 63	May 1964	+	+			Pr, 4 weeks	6-MP	0	6 months
01 LA 63	1st, Nov. 1964	+	+			Pr, V, week	6-MP	0	4 months
03 LA 65	2nd, March 1965		+	+		Pr, V, MTX (IR), week	6-MP+MTX	Pr, V once	4 months
03 LA 65	3rd, August 1965		+	+		Pr, V, MTX (IR), week	6-MP+MTX	Pr, V once	5 months
Phase I	4th, Nov. 1965		+			Ara-C 30 mg/daily	Ara-C twice weekly, 20 mg SC	0	2.5 months
Phase I	5th, March 1966		+			RU 20 mg/daily for 6 days	RU 20 mg/weekly	0	6 weeks
	6th, May 1966	+	+			RU 40 mg/daily for 5 days	Aplasia: died May 1966		

Acute leukemia: lymphoblastic—myeloblastic—undifferentiated—promyelocytic—monoblastic.
Clinical manifestations: Cl oneal; He matologic; Me ningeal; Tu moral.
Treatment employed: Prednisone: Pr; Methotrexate: MTX; Vincristine: V; 6-mercapto-
purine: 6-MP; cytosine arabinoside: Ara-C; rubidomycin: RU.

days). But the marrow aplasia that followed was rapidly fatal, death occurring from septicemia. The first phase of aplasia, lasting 15 days, was clinically well supported, there being no temperature and no focus of infection.

POI . . . Jean-Pierre, aged 13 years, had an acute lymphoblastic leukemia that was treated in September 1965 with a combination of prednisone and vincristine for six weeks (protocol 02 LA 66). He was then on maintenance treatment for four months with 6-mercaptopurine and methotrexate, with an interruption for reinduction with prednisone and vincristine in the second month. The first relapse occurred in March 1966. A fresh attempt with prednisone (5 mg/kg daily) and vincristine (3 mg/kg weekly) met with failure after four weeks, the cell-rich marrow showing 100% hemoblasts.

With a daily dose of 1 mg/kg of rubidomycin for five days complete remission was obtained in ten days after a phase of aplasia (800 white cells) that was well tolerated clinically. Fifteen days after the last injection of rubidomycin the marrow was normal and the blood showed 12.5 g hemoglobin, 2% of reticulocytes, 240 000 platelets, and 3200 white cells, with 62% polymorphonuclear neutrophils.

Maintenance treatment with rubidomycin, in a dose of 1 mg/kg weekly, was continued for five weeks. The relapse that followed was sharp, both as it affected the blood and clinically—the liver and the spleen were enormous—, but it responded again to rubidomycin in a dose of 2 mg/kg daily for five days. However, the aplasia that followed was extremely severe and complicated by septicemia caused by *Klebsiella pneumoniae*, which rapidly brought about the patient's death.

5. Overall Results. Effect of Rubidomycin Alone on the Course of Acute Lymphoblastic Leukemia

By analysing this first series of cases and literature we collected later, and by comparing our personal observations with those published by other authors [114, 167, 171, 214], we are in a position to make the following comments:

1. Rubidomycin can often bring about complete remission in acute lymphoblastic leukemia, the proportion of complete remissions being of the order of 60% for forms of the disease treated with it immediately either on their first appearance or at later relapses. When other drugs have failed, the rate of success with rubidomycin is still 45%. These results differ appreciably from those obtained by American authors using daunomycin in the same conditions. C. TAN especially [210, 212, 213] obtained apparently less clear-cut results on the same daily dose but for a shorter period (2—3 days): out of 16 cases treated, one complete remission, eight partial remissions of good quality (M1—H2—P1—S1), and two partial of poor quality. Of these 16 cases, however, four out of the five not previously treated had a remission. In our study we treated only three patients at the first manifestation of the disease. One had a remission after two injections of 1 mg/kg five days apart, receiving in all 2 mg/kg; while for the other two, two courses of five days on 1 mg/kg were required before a good remission was obtained and maintenance treatment could be instituted.

MATHE et al. [172], HARDISTY [114], HOLLAND [123] and MASSIMO [168, 169, 170] in acute lymphoblastic leukemia obtained only about three remissions in 10 cases. It seems to us that the patients in their cases were treated intermittently with several courses and that the subsequent late aplasia was more severe, the marrow regenerating imperfectly as a result. It is difficult at present to say whether these discrepancies in findings are due to the fact that in most of the published reports the series were insufficiently large or whether the reason is difference in methods such as different dosages, longer or shorter periods of administration, or continuous or inter-mittent injections.

2. There is no cross-resistance between the various drugs used for acute lympho-blastic leukemia (cortisone, vincristine, methotrexate, 6-mercaptopurine, and cytosine arabinoside) and rubidomycin. As our first series shows, rubidomycin can bring about complete remission when the disease is refractory to other forms of treatment.

3. The action of rubidomycin is often very rapid. The first changes occur at the earliest after the first injection, at the latest on the fifth or sixth day. The average time needed to obtain complete remission is 20 days, the range being from 10 days to 40 days. This scatter is essentially because of the varying duration of the period of aplasia in different patients and of the time taken for the marrow to regenerate.

4. Rubidomycin causes very remarkable morphological changes in the lympho-blasts (page 144). Thus a careful cytological examination of the bone marrow picture helps in assessment of the effect of the treatment. The appearance of giant cells a little before the period of aplasia sets in is an indication for stopping treatment with rubidomycin.

5. The duration of the complete remission caused by rubidomycin ranges from 15 to 250 days [132, 135] when treatment is continued during the remission (page 108).

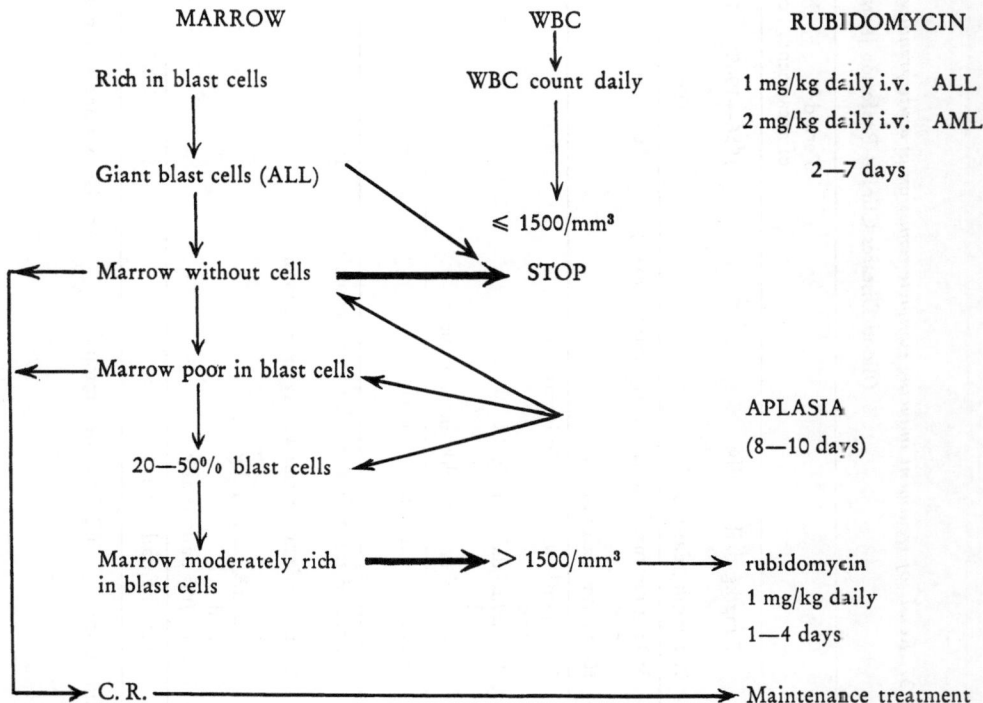

Rubidomycin injections should be stopped before the white cell count drops below 1500 per mm³ if the peripheral count falls rapidly. They can be resumed before the count begins clearly to rise when bone marrow aplasia has not been obtained with a first course of rubidomycin.

Fig. 51. Acute leukemia. Schedule of attack treatment with rubidomycin employed alone. ALL: acute lymphoblastic leukemia. AML: acute myeloblastic leukemia

Table 37. *Result of treatment inducing complete remission in acute granulocytic leukemia in terms of treatment employed*
(Blood Diseases Clinic—Professor Jean Bernard)

Form of treatment	Period of treatment	No. of cases	% complete remissions	References
Prednisone 3 mg/kg daily orally and MTX 0.15 mg/kg daily orally	1956—1960	118	12.5	Jean Bernard, M. Boiron et al. 1964 [23]
Methyl-GAG 350 mg/m² three times weekly intravenously and 6 MP 100 mg/m² daily orally	1963	45	33	M. Boiron, Cl. Jacquillat et al. 1965 [35]
Methyl-GAG 350 mg/m² twice weekly intravenously and 6 MP 90 mg/m² daily orally and MTX 15 mg/m² twice weekly intravenously	1965	24	35	Results not published
Ara-C 30 mg/m² daily intravenously	1965	39	28	Jean Bernard, M. Boiron et al. 1966 [20]
Ara-C 30 mg/m² three times weekly intravenously and Methyl-GAG 350 mg/cm² three times weekly intravenously and 6 MP 90 mg/m² daily orally and Prednisone 40 mg/m² daily orally	1966	32	43	M. Weil, Cl. Jacquillat [230]
Rubidomycin 2 mg/kg daily intravenously	1966—1967	64	54	Present study

MTX: methotrexate   6 MP: 6-mercaptopurine   Ara-C: cytosine arabinoside

6. Rubidomycin with great frequency causes profound aplasia, and treatment must be stopped according to the information furnished by blood and marrow examination; the white cells must be counted daily and a marrow examination carried out twice or three times weekly during the attack treatment. We consequently propose at present the following *schedule of treatment*. Rubidomycin should be injected daily at a dose of 1 mg/kg, and the treatment should be stopped when the white cell count falls to 1500 per mm³. If the leukopenia grows worse after the cessation of treatment and the marrow blast cells change, blood regeneration should be awaited before initiating marrow control treatment. If on the other hand the leukopenia does not grow worse and the marrow remains rich in blast cells, a second course of treatment should be started. An impoverished marrow with blast cells often calls for one or two additional injections at one or two days' interval, but only when the peripheral blood white cell count is over 1500 per mm³ on two successive examinations (Fig. 51). The interpretation of residual blastosis in an impoverished or very impoverished marrow is always difficult (Fig. 33). When the treatment of patients with leukopenia is being carried out, it is only by marrow examination repeated at very frequent intervals that treatment can be guided. In forms of the disease with excess of white cells and in some that are very sensitive to rubidomycin, the fall in the number of white cells may be extremely rapid (2—3 days), and the slope of the curve then indicates when treatment should be stopped, which should be earlier than in other cases, two days at least before the white cell count reaches 1000—1500 per mm³.

7. Because of the danger to the heart of a total dose of over 25—30 mg/kg, this is the limit to which the drug can be used in attack treatment or in reinduction.

In spite of the risk to the heart, it can be stated that rubidomycin is a major drug in the treatment of acute lymphoblastic leukemias (Table 44).

Chapter 9

Acute Granulocytic Leukemia
Acute Myeloblastic and Promyelocytic Leukemia

1. Preliminary Observations

It should be remembered that acute leukemias of the granulocytic series are of extreme and persisting severity and that therapeutic progress has been very slow [91]. Up to 1960 the percentage of complete remissions with purine and folic acid antagonists did not exceed 12⁰/o in a series of 118 cases observed between 1956 and 1960 [23, 24]. The situation improved a little with methyl-GAG [19] and then with cytosine arabinoside; but the rate for complete remissions hardly rose above 25⁰/o with cytosine arabinoside used by itself [20, 83], 33⁰/o with the combination of methyl-GAG and 6-mercaptopurine [19, 35], and 43⁰/o with the combination of 6-mercaptopurine, cytosine arabinoside, methyl-GAG, and prednisone (32 cases) [230]. Rubidomycin was the first drug that, when used alone, gave a percentage of complete remissions slightly above 50⁰/o (in a matching series of 64 cases) [36] (Table 37).

These complete remissions were obtained not only with new patients with acute myeloblastic leukemia treated from the outset but also with patients seen later, either after other attempts at treatment had failed or on relapse when remission had been obtained previously with other drugs. As Table 38 shows, they were obtained both in children and in adults.

Table 38. *Cases [64] of acute granulocytic leukemia treated with rubidomycin*

Cytological type		Age of patients	
Myeloblastic AL	59 cases	Adult granulocytic AL	48 cases
Promyelocytic AL	5 cases	Children's granulocytic AL	16 cases

2. Dosage

The dose should be high—between 1.5 and 2 mg/kg daily, or 45—100 mg/m², and the drug should at first be administered daily. The total dose should vary from 3 to 22.5 mg/kg, with an average of 12 mg/kg. Thus a course should consist of 3—6 consecutive days of treatment, with the one or two supplementary injections sometimes needed being given at another time. During the attack treatment, the blood should be examined every other day and the marrow twice or three times weekly.

3. Study of a First Personal Series [36, 78]

a) Patients Treated

The number of cases was 64, distributed as shown in Table 38.

Among these cases 43 were treated at the first manifestation of the disease. The great majority had never had any previous treatment, but four had received prednisone, 6-mercaptopurine, or methotrexate alone or in various combinations for 4—6 weeks without any result.

Twenty-one patients were treated on relapse, 13 on first relapse, six on second, and two on third. The treatment they had received previously at the start of the disease or on relapse had included cytosine arabinoside, methyl-GAG, prednisone, and methotrexate, separately or combined in various ways according to the stage of treatment (attack or maintenance).

Course of the disease. Analysis of the therapeutic effect. The important fact is that complete remission could be obtained in about half of the cases. Table 39 sums up our experience at present: if the disease was treated at its onset, complete remission was obtained in 26 out of 43 cases, i. e. in 60%. In children it was obtained in 11 cases out of 16.

Frequency of remission. The remission rate obtained by other teams of workers is not as high [197]. While it is common in drug treatment, at least initially, for different centers to obtain different results or for the same team to do so at different times, it seems to us that the discrepancies in this case, as we said above, were due to a dosage that was often inadequate [57]. As we shall have occasion to say later, the conduct of treatment with rubidomycin is often far from simple (page 116).

Table 39. *Results obtained with rubidomycin in induction treatment for acute granulocytic leukemia (in terms of the stage of the disease and the patient's age)*

	1st attack		1st relapse		2nd relapse		3rd relapse		Total
	Children	Adults	Children	Adults	Children	Adults	Children	Adults	
Complete remissions	6	20	3	3	1	1	1		35
Incomplete remissions		4	1	1		1	1		8
Failures									
Total resistance			1	1					2
Aplasia with blast cells	1	5		1		1			3
Aplasia without blast cells		7		3	1				11
Total	7	36	4	9	3	3	2		64

Complete remission in 35 cases (54⁰/₀) was obtained after a variable time, averaging 31 days after the first injection of rubidomycin, with a minimum of 15 days and a maximum of 62. The total doses of rubidomycin varied from 3 mg/kg to 38 mg/kg with an average of 12 mg/kg, but the scatter was great (Fig. 52).

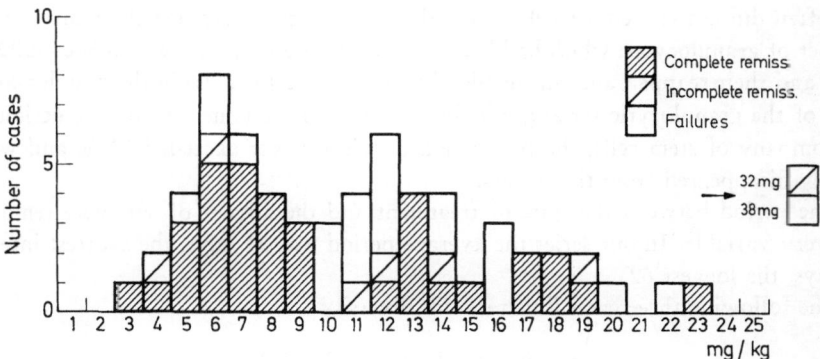

Fig. 52. Total dose of rubidomycin used on 64 patients with acute granulocytic leukemia. Results of 1—3 courses of treatment

Complete remission was always preceded by a phase of severe aplasia, which started on the sixth day at the minimum and the 21st at the maximum, but generally around the seventh day.

The appearance and intensity of the aplasia naturally made adjustment of the treatment necessary. Put in a nutshell, the attitude we adopted was determined by daily—or at least very frequent—consideration of three factors: the number of peripheral leukocytes, the percentage of marrow blast cells, and the richness of the marrow in cells on smears.

Treatment should be stopped when the white cell count is in the neighbourhood of 2000 per mm³ if the fall in their number is very rapid, or 1500 if the fall is slower (cf. Fig. 51, p. 111). If the bone marrow is still blastic at that stage and of normal or average cell richness, fresh injections could be made if the white cell count is over 1500 per mm³. If the marrow is poor in cells and the leukopenia persists, no new injections should be made, but they could be resumed when the cell count rises and marrow blastosis continues. In this case, when the marrow smear is poor in cells but still contains myeloblasts in a proportion of more than 30% after two successive examinations, a fresh injection can be considered with a dose of 1 mg/kg, even if the white cell count is less than 1000 per mm³, for it is no easy task to determine the beginning of complete remission in a very cell-poor marrow. It seems unnecessary to require all the classical criteria for complete remission (page 102) at that stage, since the bone marrow and the blood do not present a perfectly normal appearance until a fairly long time after the disappearance of the leukemic cells.

We pointed out above the importance and difficulty of the problems involved in the interpretation of certain blastoses at the stage of regeneration after aplasia. Careful cytological observation of the blast cells and their cytochemical characteristics and development make it reasonably often possible to differentiate regeneration blastosis from a relapse or an incomplete result. This distinction is important because the conduct of the treatment is the exact opposite in the two cases.

b) Time of Complete Remission

In all the cases that went into remission, the duration of the aplasia and the high temperature that was associated with it towards the eighth day did not exceed 15 days and often did not exceed a week. The fall in temperature heralded the remission, the number of granulocytes, which had been close to 0, rose progressively to 300—500 per mm³, and their reappearance in the blood was signalled by a rise in the platelet count. Cells of the granulocytic series gradually returned to the bone marrow, sometimes in the company of stem cells, the proportion of which never exceeded 10%, and which rapidly disappeared from the smears.

The period between the start of treatment and the advent of complete remission was very variable. In our series the average period was 31 days, the shortest interval 15 days, the longest 62.

The following three cases illustrate this diversity.

α) *Average Interval* (Fig. 33)

LON . . . Monique, aged 14 years. This patient had an acute myeloblastic leukemia, discovered through a fulminating spasmodic paraplegia appearing after several weeks of cord pains and sensory disturbances of the upper limbs.

She was hospitalized on 21 January 1967, and the blood showed 30% of hemoblasts, the bone marrow 70% (blast cells with Auer bodies). Before her transfer to a neurosurgical unit, she was given two injections of rubidomycin in two days, of 3 and 2 mg/kg. Operation completely removed a myeloblastic tumor, hard and poorly vascularized, situated between D 1 and D 4 posteriorly to the dura mater. With two fresh injections of rubidomycin of 2 mg/kg she obtained a complete remission after ten days of blood and marrow aplasia (marrow without cells; 800 white cells per mm³). She had a relapse after seven months of complete remission without making any marked neurological recovery.

β) *Short Interval*

PRO . . . Michele, aged 69 years. This patient had been tired and had lost weight for a month, and in September 1967 she was found to have acute myeloblastic leukemia. Clinical examination revealed only a moderately enlarged spleen, but the blood examination showed

3 250 000 red blood cells, 20 000 platelets, and 245 000 white cells, of which 95³/o were hemoblasts. The bone marrow was rich in cells, 100⁰/o myeloblastic.

Rubidomycin was prescribed at a dose of 2 mg/kg. Two injections were given, for the following day the white cell count was 70 000 per mm³, the day after 5000, and the fourth day 1000. This phase of blood and marrow hypoplasia was accompanied by septicemia and hemorrhages. On the 10th day the white cell count started rising, and on the 15tⁿ it was 4500 per mm³; the proportion of polymorphonuclears was 76⁰/o, of lymphocytes 21⁰/o, and of monocytes 3⁰/o; the platelet count was 160 000. The bone marrow was normal in cell content and there were no abnormal elements; 48⁰/o of the cells were of the granulocytic series, 43⁰/o of the red blood cell series, and 7⁰/o lymphocytes, while the megakaryocytes were normal. Complete clinical and hematological remission was obtained.

Two other similar observations are described on pages 96 and 122. In the former there were cardiac complications, but after five days of treatment the ensuing aplasia lasted for only four days before remission. In the second complete remission was obtained after 3 mg/kg of rubidomycin.

γ) Long Interval

GAY ... Bernard, aged 21 years (Fig. 54). This was a case of acute myeloblastic leukemia in a young man aged 21 years. The first sign of the disease was a serious attack of stomatitis in May 1966. The patient was hospitalized on 25 May, and acute leukemia was diagnosed.

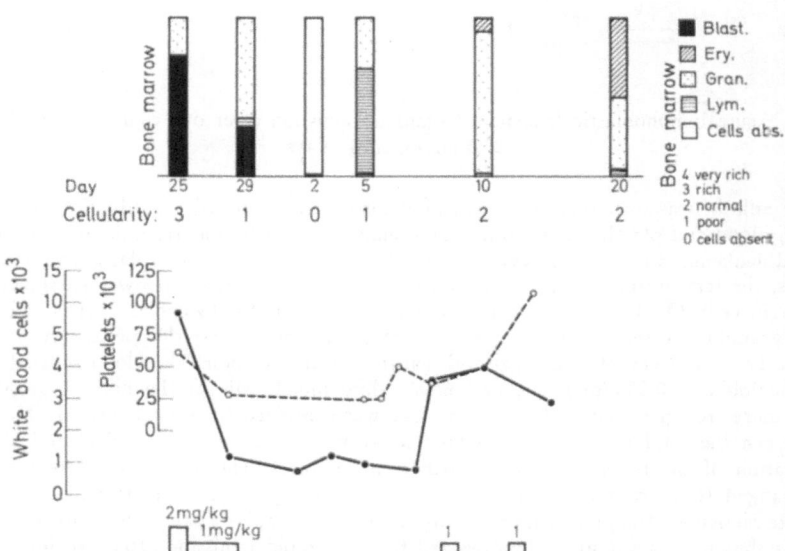

Fig. 53. Acute myeloblastic leukemia. Complete remission after 6 mg/kg of rubidomycin administered in five days

The blood showed 3 740 000 red blood cells, 80 000 platelets, and 18 000 white blood cells, of which 13⁰/o were polymorphonuclears and 66⁰/o hemoblasts of the granulocytic series with Auer bodies. The marrow was very rich in cells, and showed 94⁰/o of hemoblasts with Auer bodies (the peroxidase test was highly positive). Clinical examination showed painful sub-maxillary adenopathies, which were probably connected with the stomatitis, and painless bilateral inguinal lymph nodes of small size; neither the liver nor the spleen was palpable.

Rubidomycin was prescribed, and the patient was given seven injections of 1 5 mg/kg, followed by four further injections of 3 mg/kg. There was a gradual disappearance of the blast cells from the marrow, accompanied by the development of marrow aplasia. The marrow became completely denuded of cells, and septicemia appeared, accompanied by a very high temperature. The peripheral blood showed 400 white blood cells. The patient suffered particularly from mouth ulcers and stomatitis; this last was extremely painful, and injections of opiates had to be given.

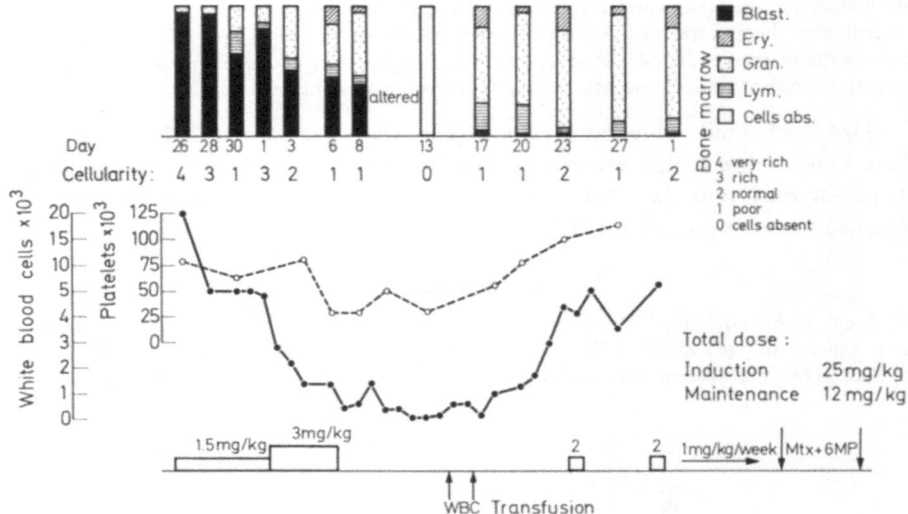

Fig. 54. Acute lymphoblastic leukemia. Complete remission after one course of rubidomycin, 22.5 mg/kg in 11 days

The aplasia was overcome by means of vigorous treatment with antibiotics, transfusions of fresh blood and platelet concentrates, and leukocyte transfusions from donors with chronic myeloid leukemia in the myelocytic stage. On 15 June, after the aplasia had lasted for 38 days, the temperature returned to normal; examination of the marrow then showed it to be poor in cells. On 1 July, however, it was cell-rich, with 2% hemoblasts, 41% of elements of the granulocytic series and 43% of elements of the red series; the megakaryocytes were normal. There was complete clinical and hematological remission. The blood showed 11.50 g of hemoglobin, 450 000 platelets, and 6600 white blood cells in the normal proportions. Maintenance treatment with rubidomycin alone was continued for four months. In November 1966, when the total dose given amounted to 41 mg/kg, clinical, radiological and electrical examination of the heart showed no cardiac disturbances. The maintenance treatment was then changed (6-mercaptopurine and methotrexate). Relapse occurred after nine months of complete remission. The patient then rapidly succumbed to aplasia caused by cytoxan.

It is thus necessary both to be prepared for very rapid remission after two injections of rubidomycin, and to be able to wait patiently for two months when remission is slow to occur, while not failing to give one or two additional injections at the right time.

c) Duration of Complete Remission

The 35 cases in which complete remission was obtained comprise:

a) four patients who died in remission, but whose death was not due to any hematological cause (the first died from a subphrenic abscess on the 35th day of complete remission, the second from cardiac insufficiency in the 5th month of remission, the third from sudden collapse on the 90th day, and the fourth from anuria on the 52nd day of remission);

b) 26 patients who relapsed after an average period of 155 days, the minimum being 30 days and the maximum 387 days;

c) 5 patients who are still in complete remission, which has lasted for more than 12, 19, 20, 22, 24 months respectively.

All the patients in complete remission have been given maintenance treatment.

The first three patients in the series were given 2 mg/kg of rubidomycin once a week. Of these, the first had a relapse after three months, at which time he had received a total cumulative dose of 31 mg/kg, and had not shown any cardiac disturbance. The second presented with a severe and fatal cardiac insufficiency (see Obs. No. 4 on p. 96) after five months of treatment, having been given a total dose of 40 mg/kg. The third was treated with rubidomycin for four months (total dose: 41 mg/kg) without any cardiac mishap. The type of maintenance treatment was then changed (methotrexate and purinethol). A relapse occurred in the tenth month, but no cardiac disturbances could be detected clinically, radiologically or electrically.

Rubidomycin was not used in the maintenance treatment of the 32 remaining patients in complete remission because of the risk of cardiac accidents. Instead, 6-mercaptopurine was given orally (90 mg/m² per day) in combination with intramuscular injections of methotrexate (15 mg/m² once a week). This maintenance treatment was briefly interrupted for a period of one week in the 1st, 2nd, 4th, 7th and 11th month for reinduction with rubidomycin (two intravenous injections of 1 mg/kg or 30 mg/m² on the first and eight days) and methyl glyoxal (two injections of 250 mg/m² on the third and fifth days). This represents an attempt to apply to acute granulocytic leukemias the principles that appear to give good results in the treatment of acute lymphoblastic leukemia (see Table 40).

Table 40. *Treatment of acute myeloblastic leukemia with rubidomycin*
Phase I (Protocol RU, October 1966)

Induction	Maintenance	Reinductions 1, 2, 4, 7, 11 months . . .
Rubidomycin 2 mg/kg daily intravenously (60 mg/kg) 2—7 days	6-mercaptopurine 90 mg/m² daily orally	Rubidomycin 1 mg/kg on days 1 and 8 (30 mg/m² intravenously)
	Methotrexate 15 mg/m² weekly intramuscularly	Methyl-GAG 250 mg/m² intramuscularly on days 3 and 5

d) Incomplete Remissions

The eight cases of incomplete remission include three in which only a partial reduction in blastosis (less than 30%) was achieved, and five in which the disappearance of the marrow blast cells was only temporary and lasted for periods of 7 to 21 days.

e) Failures

Failures are regarded as those cases in which a satisfactory remission is not achieved and in which the disease takes a fatal course. They are of several types. The frequency with which these different types occurred in our personal series is shown in Table 31, p. 88.

1. Death may be due to an accident, such as a cerebro-meningeal hemorrhage, or more often an extremely acute septicemia, which carries the patient off during the first few days of treatment and before this has time to have any effect (four cases).

2. Death may be due to aplasia after the disappearance of the blast cells. Death from non-blastic aplasia has unfortunately occurred fairly frequently. It must be expected and feared as the outcome of aplasia when this has proved impossible to overcome. The risk of aplasia during the first week of treatment may be reduced by the more vigorous use of the techniques of hematological resuscitation, but this is far from making it possible to avoid all risk of accidents later on.

3. The situation is more often rather difficult to assess. A marked partial reduction of marrow blastosis may have been achieved by the use of rubidomycin, but at the price of serious peripheral aplasia. The clinical picture is one of aplasia with partial marrow blastosis. If the treatment is stopped, the proportion of marrow blast cells immediately increases; if the treatment is continued, the aplasia becomes more serious, and the patient dies of a combination of acute aplasia and only partially controlled leukemia, without the correct treatment having been discovered. It is thus frequently found that the treatment achieves only a partial decrease in the number of blast cells.

4. Complete failure, i. e. complete resistance to the treatment and persistence of the blastosis unchanged is, on the other hand, very rare. Only two cases of such failure were found in our series.

4. General Survey. General Features of the Course of the Disease. Prognosis

An examination of the results obtained shows that the cases can be divided into three groups:
1. cases in which the treatment quickly gives an improvement;
2. cases in which the disease continues unaffected to its final outcome;
3. cases in which the course of the disease is uncertain.

Most cases fall into the third group. There is at times a tendency towards remission after a period of agonizing uncertainty, at times a tendency towards a fatal outcome after hopes have been raised for a time by an improvement in the patient's condition.

The factors on which any attempt to forecast the course of the disease and, in general, any attempt at prognosis, must be based include the patient himself, the type of leukemia concerned, the hospital organization and the experience of the hematologist. Precise statistical comparisons are not possible in this study, because of the relatively small number of patients involved; we shall therefore make only the following comments.

a) Age

It is generally accepted that, in children, acute granulocytic leukemia is a slightly less serious disease and one in which remissions occur a little more frequently than in adults. In the case of rubidomycin, no precise conclusions can be drawn from our personal observations. With patients less than 30 years of age, however, complete remissions were obtained in two thirds of the cases, or 17 out of 26 (see Table 41).

Table 41. *Results obtained with rubidomycin in terms of the patient's age*
(acute granulocytic leukemia)

	Complete remissions	Incomplete remissions	Failures	Total
0—1 year				
2—5 years	3	1		4
6—10 years	3	1	3	7
11—20 years	5	3		8
21—30 years	6		1	7
31—40 years	3		6	9
41—50 years	6	1	1	8
51—60 years	3	1	3	7
61—70 years	4	1	7	12
over 70 years	2			2
Total	35	8	21	64

b) Type of Leukemia

Earlier work, in particular by our group [23] has shown that the prognosis in acute myeloblastic leukemia depends on three factors.

Platelets. The maintenance of a normal, or slightly reduced, platelet count is generally a factor indicating a favorable prognosis. The data obtained with treatment with rubidomycin are shown in Table 42.

Leukocytes. It is known that an initial high leukocytosis indicates an unfavorable prognosis in acute lymphoblastic leukemia; remission is less frequently achieved, more difficult to achieve and of shorter duration. No conclusions can be drawn from Table 43 in this connexion. It will be noted, however, to the credit of rubidomycin,

Table 42. *Results obtained with rubidomycin in terms of the initial platelet count*
(acute granulocytic leukemia)

Platelet count	Complete remissions	Incomplete remissions	Failures	Total
100 000 per mm³ or over	16	2	5	23
50 000—100 000 per mm³	14	3	10	27
Less than 50 000 per mm³	5	3	6	14
Total	35	8	21	64

Table 43. *Results obtained with rubidomycin in terms of the initial white blood cell count*
(acute granulocytic leukemia)

No. of white blood cells per mm³	Complete remissions	Incomplete remissions	Failures	Total
Over 100 000 per mm³	2	1		3
50 000—100 000 per mm³	2	1	4	7
Less than 50 000 per mm³	31	6	17	54
Total	35	8	21	64

that very good remissions can be obtained even in cases with very marked hyperleukocytosis. In one case, for example, marrow aplasia was complete after administration of 3 mg/kg of rubidomycin; the blood initially showed 160 000 myeloblasts
per mm³ and the marrow had been completely invaded. After eight days, the blood
showed only 60 leukocytes per mm³. The marrow had returned to normal by the
18th day, and normal platelets, shortly afterwards followed by polymorphonuclears,
reappeared in the blood. In this case, therefore, complete remission was obtained by
the administration of a small dose of the drug, the second injection of rubidomycin
being only 1 mg/kg in view of the rapid reduction in the leukocytosis and the high
sensitivity to the drug of the blast cells in this patient.

Highly "tumoral" forms, in which there are voluminous adenopathies, splenomegaly and hepatomegaly, accompanied by a moderate degree of leukocytosis, must
be classified with the markedly hyperleukocytic forms.

The precautions mentioned on p. 78 must be borne in mind in both cases: plenty
to drink and the immediate administration of Allopurinol, so that accidents due to
the destruction of large numbers of cells and the disadvantages consequent on a
delay in the application of urgently needed treatment can both be avoided.

Extent of invasion of the marrow. There are a great number of forms of the
disease with partial invasion of the marrow. They may be conveniently be divided
into two classes:

1. Partial invasion of the marrow by blast cells with the marrow rich in cells
(under 50%). The prognosis is relatively favorable. Remissions are very common and
are often rapidly obtained; two complete remissions without dangerous aplasia were
obtained in patients in first relapse, which was detected by the bone marrow picture
when the clinical picture and the blood count were still normal.

2. Partial invasion of the marrow with the marrow poor in cells. These forms are
very uncertain in their course. We have not yet carried out trials on a large enough
scale to be able to reach any conclusions.

c) Hospital Organization

The prognosis must largely depend, throughout the struggle with the disease, on
the standard of symptomatic treatment provided. This, in turn, depends on a large
number of factors, such as the hospital structure, the training of the nursing staff, the
way in which the transfusions of platelets and white blood cells are organized, and
the treatment with antibiotics.

Even where conditions appear to be satisfactory, the patient's fate may nevertheless depend on chance circumstances, such as the availability, at the right time, of
a donor suffering from chronic myeloid leukemia in the myelocytic stage, or the
occurrence of an epidemic of *B. pyocyaneus* infection, which carries off patients in
the aplastic state in whose case regeneration of the marrow could otherwise reasonably have been expected.

d) Experience of the Hematologist

This would hardly seem to be worth mentioning but for the fact that it does
explain, at least in part, the differences between certain of the results obtained.
Rubidomycin has been used in therapy for too short a time, even although several
hundred patients have already been treated, for it to be possible to lay down any

precise rules. The experience of the hematologist will therefore be a factor of major importance for a number of years to come. We can, nevertheless, recommend the same schedule of treatment as that used for acute lymphoblastic leukemia (see Fig. 51), except that the daily dose of rubidomycin is 2 mg/kg instead of 1 mg/kg.

e) Comparison of our Personal Series with the Results Obtained by other Workers

As things are at present, this comparison can only be somewhat incomplete; it is based on what are, at times, somewhat imprecise oral communications rather than on the inadequate number of published papers.

Although incomplete, the comparison does bring out certain undeniable differences between series such as ours, in which relatively successful results were obtained, and others in which the results were less successful. It is important to try and find the reasons for these differences, of which some will be common to all forms of treatment with antileukemic drugs, while some will be peculiar to rubidomycin.

α) General Causes

1. It can almost be stated as a law that the results obtained in treatment with antileukemic drugs by teams used to working together are better than those obtained by teams which deal only with a small number of cases. This difference has been pointed out in the comparisons made in recent years of the results obtained by various teams, both for VAMP and for cytosine arabinoside.

2. The dose is an important factor. Certain teams rapidly find the most suitable dose, while others have greater difficulty in doing so; their approach is more hesitant in character, and they sometimes become discouraged. It is clear that, with rubidomycin, the best results have been obtained by teams which have used high doses, although such doses lead to aplasia with its attendant dangers.

β) Special Difficulties

It is fairly easy, at the present time, to decide on the best initial dose, but it is still difficult to lay down precise rules as to the way in which the attack treatment should be continued. As already pointed out, the correct path must be found in each case between overdosage and doses which are inadequate. It is not difficult to see that, under these conditions, large differences between various series of results are to be expected. Two comments would seem to be in order at this point:

1. Some risks can be accepted in the treatment of the most serious forms of leukemia.

2. The proportion of complete remissions (50% in our series) should be in the range 35—55%, if the dosage is correct. As mentioned above, this is higher than that obtained with any other drug for the treatment of acute granulocytic leukemia, when used alone.

5. General Indications

It would be premature, at the present time, to attempt to reach any definite conclusions as to the effectiveness of rubidomycin in the treatment of acute granulocytic leukemia. It will nevertheless be of value to compare the results obtained with those obtained by means of other forms of therapy, as shown in Table 37.

It is clear that the most interesting comparison would be that between the results obtained by the same team using different methods.

It so happens that at the time that rubidomycin appeared, a trial carried out by our group [230] had shown that the complete remission rate could be markedly increased by the so-called quadruple method, in which cytosine arabinoside, methylglyoxal bisguanylhydrazone, 6-mercaptopurine and prednisone were used in combination.

Thus, as things are at present, rubidomycin would seem to be indicated, in preference to other drugs, for the treatment of acute granulocytic leukemia.

The initial combination of rubidomycin with the four drugs used in the quadruple treatment, or with some of them, should be considered, and this question will be referred to again later.

6. Acute Promyelocytic Leukemias

Acute promyelocytic leukemia calls for special mention; we pointed out in 1959 that this disease was extremely serious in character and that it possessed special clinical features [17, 22, 29]. Complete remissions can be produced by rubidomycin in acute promyelocytic leukemia. Our personal series included five cases and three complete remissions were obtained by the treatment. The following case may be given as an example.

CAR . . . Michèle, aged 23 years.

This patient was hospitalized on 25 November 1966, and acute promyelocytic leukemia was diagnosed.

Clinically, the onset of the disease would seem to date back to May 1966, when the patient noted marked debility accompanied by episodes of fever with temperatures of 40° C; these were treated by means of antibiotics. The debility lasted for three months.

From the beginning of November 1966, multiple spontaneous ecchymoses appeared. On 25 November there were 73% of promyelocytes in a very cell-rich marrow. The blood showed 3 130 000 red blood cells, 10 000 platelets and 1000 white blood cells, of which 20% were hemoblasts and 26% promyelocytes. No hypertrophy of the hematopoietic organs could be detected by clinical examination, and liver, spleen and lymphnodes were normal; on the other hand, the skin showed numerous ecchymoses forming large plaques. A check on the blood-clotting mechanism showed a hypofibrinogenemia of 1 g%.

Treatment with rubidomycin was begun, and consisted of two courses of the drug, the first of 6 mg/kg in three days, and the second, after an interval of eight days, of 6 mg/kg in eight days.

The general condition of the patient had undergone a complete transformation by the time the first course of rubidomycin had been completed; there were no new ecchymoses and no hemorrhagic syndrome and the fibrinogen content of the blood was 2.25 g%. On 8 December, before the second course of treatment with rubidomycin was begun, the marrow was relatively poor in cells and contained 51% of hemoblasts; both hemoblasts and promyelocytes had disappeared from the peripheral blood. A complete remission was obtained by means of the second course of rubidomycin, after a short period of marrow hypoplasia, without the appearance of any clinical signs of septicemia and without a high temperature. On 28 December, the marrow was normal, with 3% of hemoblasts, 51% of elements of the granulocytic series and 46% of elements of the red series. The blood showed 12 g of hemoglobin, 280 000 platelets and 4 400 white blood cells, with a normal blood picture.

Treatment with 6-mercaptopurine and methotrexate was begun. It was interrupted in the 1st, 2nd, 4th, 7th, 11th, 16th and 22th month for reinduction, over a period of a week, with rubidomycin and methyl-GAG.

The clinical and hematological condition of the patient was perfectly normal after 26 months of complete remission, and she has returned to her university studies.

7. Acute Monoblastic Leukemia

We have little experience in this field. The position is complicated because of the difficulty of distinguishing between the various forms of the disease. The majority of the cases are of acute myeloblastic leukemia, but sometimes there are also cases of true leukemia with cells of the reticulo-histiocyte series. In such cases, it would seem possible for complete remissions in acute leukemia to be obtained with rubidomycin. Thus we have obtained one complete remission out of a group of five patients treated with the drug; of the remaining four cases, three showed partial sensitivity, but it proved impossible to control the aplasia, while the fourth case was one of total resistance to rubidomycin. It would appear that, in these forms of the disease, the malignant cell population consists of cells of two types; of these, the myeloblastic type is very sensitive to rubidomycin, while the reticulo-histiocyte type is only partially sensitive to the drug.

Chapter 10

Combination Therapy in the Treatment of Acute Leukemia

1. General

The use of sequential or combination therapy in the treatment of acute leukemia was proposed by one of us in 1952 in the case of aminopterine and cortisone [28]. The value of such treatment was questioned for a time, but it was later generally accepted and is now widely used. This wide usage and the fact that the therapy is almost universally accepted are the consequence both of experimental evidence of its effectiveness and the hope that even better results will be obtained in the future. Thus the remarkable work of SKIPPER [205] and GOLDIN [106] has shown that, for mice inoculated with leukemia L 1210, there is a relation between the number of cells injected on the one hand, and the duration and severity of the leukemia on the other [107, 204]. This would seem to justify the hope that a similar relation exists for spontaneous leukemia, and especially for human leukemia. Then there is the fact that there is no cross-resistance between the different antileukemic drugs used when they are of different chemical type, the possibility that the effects of such drugs used simultaneously may be the sum of their effects when used singly or that potentiation may occur, the fact that the number of complete remissions increases when a number of drugs are administered simultaneously (see Table 44), and the hope that both the length of the first remission and the life of leukemic patients may be prolonged by combination therapy [18, 88, 89].

Table 44. *Induction of complete remission (CR) in acute lymphoblastic leukemia in children not previously treated, in relation to the treatment used (% of CR)*

Treatment	% CR		No. of patients	References
	One drug only	Combinations		
Methotrexate	22%		48	
		44%	42	Frei (1961) [88]
6-mercaptopurine	26%		43	
		82%	154	Frei (1965) [89]
Prednisone 1 mg/kg	57%		72	Freireich (1963) [93]
	61%		55	Jean Bernard (1962) [24]
ACTH	66%		100	Pierce (1957) [179]
		84%	63	Acute leukemia B (1965) [1]
Vincristine	57% [a]		103	Karon (1963) [141]
		88%	95	Jean Bernard (1966) [18]
Prednisone 3 mg/kg	74%		62	Jean Bernard (1962) [24]
	62% [b]		180	
		90%	130	Jean Bernard (1967) [21]
Rubidomycin	60% [b]	96% [c]	38	Jean Bernard (1966) [26]

[a] Patients in first relapse.   [b] Adults and children.   [c] children

The fact that the complete remission achieved by induction treatment with prednisone or vincristine is of short duration [141] would appear to confirm that these drugs produce only a partial reduction in the number of hemoblasts. A possible explanation of the uselessness of continuing with induction drugs is that the exponential proliferation of the leukemic cells which remain more than counterbalances the effects of the drugs, which act only on a fixed proportion of the malignant cells.

There is no doubt that the duration of complete remission can be increased by maintenance treatment. Nevertheless, while the median duration of complete remission was 7 months with 6-mercaptopurine [24, 30, 93], it still does not exceed 12 months even with intermittent methotrexate, the maintenance treatment that has given the best results so far [1, 127, 196].

The length of the period of unmaintained remission can be markedly increased by high dose combination therapy aimed at achieving the maximum initial cell kill; the slope of the remission curve provides an index of the value of the various combinations. According to Johnson and Zellen, the distribution of relapses for unmaintained remissions follows an exponential curve characterized by the guarantee time, i. e. the period during which no relapses have occurred and the occurrence of the first relapses (the distribution of relapses for maintained remissions, on the other hand, does not appear to obey this law).

Since the tolerance of the host limits the drug doses which can be used, numerous forms of combination therapy have been suggested. These can be divided essentially into three types:

1. the simultaneous administration of several highly active drugs, as in the VAMP method [96, 142];

2. the cyclic or sequential administration of a number of drugs; and

3. the systematic reinduction method, proposed by us.

a) The Simultaneous Use, or the Use at very Short Intervals, of Four Main Drugs

This was proposed by the team at the National Cancer Institute, Bethesda (FREI-REICH et al., 1964, HENDERSON, 1967) [119, 120] and is known as the VAMP method from the initial letters of the drugs used:

Vincristine	2 mg/m²/i.v. per week
Methotrexate	20 mg/m²/i.v. per day, for four days
6-Mercaptopurine	60 mg/m²/p.o. per day
Prednisone	40 mg/m²/p.o. per day.

The treatment extends over a period of ten days, and five successive courses are given at intervals of ten to 30 days, depending on the hematological tolerance. There is a greater risk of drug accidents with this method, the period between successive courses is short, and aplastic accidents are possible and call for vigorous hematological measures (platelet transfusions, leukophoresis).

The median duration of unmaintained remissions is then only four months. The so-called BIKE method gives a median remission duration of five months. This method involves two successive courses of treatment in the following sequence: methotrexate (15 mg/m² daily for five days), 6-mercaptopurine (1000 mg/m² daily for five days) and cytoxan (1000 mg/m² daily in a single injection) [95].

The median duration of unmaintained remission is six months after consolidation treatment which includes, in addition to methotrexate, 6-mercaptopurine and cytoxan, 1,3 bis(2-chloroethyl)-1-nitrosourea (BCNU) [47, 113].

b) The Sequential or Iterative Method

This is the method of ZUELZER [234], in which 6-mercaptopurine (2.5 mg/kg daily) and methotrexate (1.25—5 mg orally daily) are used alternately every three months. The median remission duration is 44 weeks. In the iterative treatment (MATHÉ et al.), each of the drugs available is administered in succession, on the principle that each antimitotic drug must, after it has been used, leave behind a small number of resistant cells, from which the malignant cell population is reconstituted. Each course of chemotherapy is separated by a period of one month, during which prednisone is given. In this way, methotrexate, a purine antagonist, vincristine, cyclophosphamide, vincaleukoblastine and methylglyoxal bis-guanylhydrazone are administered in succession. The results obtained in this way with patients treated from the first stages of the disease have been encouraging. The median duration of the first remission is 20 months (13 patients), and that of survival 26 months [173].

c) The Reinduction Method [16, 56, 129, 134]

This method is based on the working hypothesis that it is necessary to kill the leukemic cells regularly while they are still relatively few in number. This hypothesis would appear to be confirmed by the fact that there is a marked increase in the duration of remission in patients treated in this way. As a consequence, we have for some years been using the following schedule in the treatment of acute lymphoblastic leukemia:

Induction with prednisone and vincristine. We have treated 95 patients (79 children and 16 adults) with prednisone (120 mg/m² daily) and vincristine (1 mg/m² for the first week and 2 mg/m² for the following weeks). Complete remissions were obtained in 86 cases, i. e. 88%.

Consolidation treatment. We used 6-mercaptopurine (90 mg/m² orally), given as soon as the prednisone was stopped, and intramuscular methotrexate (15 mg/m²) twice a week, the first injection being given one month after the last injection of vincristine.

Maintenance treatment. This combined 6-mercaptopurine and methotrexate, at the same dosage, but injected only once a week.

Reinduction. This involves the temporary interruption of the maintenance treatment so that reinduction, once again with prednisone and vincristine, can be carried out. The patient must be hospitalized. Three intraspinal injections of methotrexate are given every two days; if there is any meningeal involvement, the injections are continued until it is controlled.

a) In a preliminary investigation, we compared the duration of the second or third remissions obtained with the various maintenance treatments used, and with the duration of the first remission. Although this investigation was not carried out with the same precision as that on the first remission, it nevertheless makes it possible to assess rapidly the effectiveness of the maintenance treatments used and the extent to which they are tolerated by the patients. Thus, in this first investigation, we carried out reinduction over a period of two weeks at intervals of 2, 3, 5, 8, and 11 months

Table 45. *Duration in months of successive complete remissions obtained in patients with acute lymphoblastic leukemia. Note the number of months on maintenance treatment involving reinductions*

Observations	Sex	Age	Duration of complete remissions in months			
			1st	2nd	3rd	4th
1	M	17	11	8	14	
2	F	8	24	1,5	35+	
3	M	12	38	7	8	
4	M	8	24	2	3	
5	M	12	10	6	1	7
6	F	7	9	2	3	
7	M	7	7	8,5	11	
8	F	6	11	6	12	
9	M	10	12	13	*12*	
10	F	7	11	3		

after complete remission. A total of ten patients were investigated, in their 2nd, 3rd or 4th relapse. The value of this method was shown by the fact that in 8 cases out of 10 (see Table 45), the duration of remission when the treatment was used was greater, sometimes markedly, than that of the previous remission, whereas the opposite was expected (the attack treatment with prednisone and vincristine and maintenance treatment with intermittent 6-mercaptopurine and methotrexate were identical in the second and third remissions, so that reinduction was used alone in the last remission). We subsequently proposed that reinduction should be carried out during the first remission, and have used two treatment schedules in succession.

b) Reinductions every six months (see Table 46, protocol 02 LA 64). An examination of the first results obtained by this method, as shown in Fig. 55, shows that the number of relapses occurring in the first six months before the first reinduction (9 out of 41 complete remissions) was particularly high between the 5th and 6th month immediately before reinduction. On the other hand, there were fewer relapses (6 out of 40) between the 6th and the 12th month. Hematological relapses are very rare after the first year; we have observed no more than six, in the 16th, 17th, 29th, 30th, 31st and 36th months respectively. The form of the remission curve, obtained by actuarial methods, provides confirmation of these results; it shows an initial rapid drop, followed by a subsequent levelling off (Fig. 56).

Table 46. *Treatment schedule for acute lymphoblastic leukemia (Protocol 02 LA 64)*

Induction		Consolidation (3 weeks)	Maintenance	Reinduction
Prednisone 100 mg/m^2 daily orally	Complete remission	MTX 15 mg/m^2 twice weekly intramuscularly (one month after last injection of vincristine)	MTX 15 mg/m^2 once weekly intramuscularly	Every six months, if remission continues: prednisone, 100 mg/m^2 daily for 15 days vincristine, on days 1, 8, 15 (1, 2, 2 mg/m^2) MTX spinally, 10, 11, 12 mg/m^2
and			and	
Vincristine 1, then 2 mg/m^2 weekly intra-venously		6-MP 90 mg/m^2 daily orally (as soon as prednisone stopped)	6-MP 90 mg/m^2 daily orally	

If the meninges are involved, spinal injections of methotrexate (MTX) are made every other day until control of the meningitis is achieved; thereafter, only once a month (MTX 10 mg/m^2).

Protocol 04 LA 65: Reinductions last only a week and involve only two injections of vincristine (1, then 2 mg/m^2). Reinductions are carried on on the 2nd, 4th, 7th, 11th, 16th, 22nd, 29th month, etc. . . . of the remission.

Spinal injections of methotrexate are started only on the 7th month unless the meninges are involved.

c) It was therefore necessary to take action at the stage corresponding to the initial slope of the curve, by increasing the number of reinductions in the early stages, after which they could be spaced out on an exponential curve matching that for the relapses. A new investigation was therefore begun (protocol 04 LA 65) in which reinduction was carried out earlier on (1, 2, 4, 7 and 11 months after remission) and over a shorter period (one week, with two injections of vincristine). As shown by

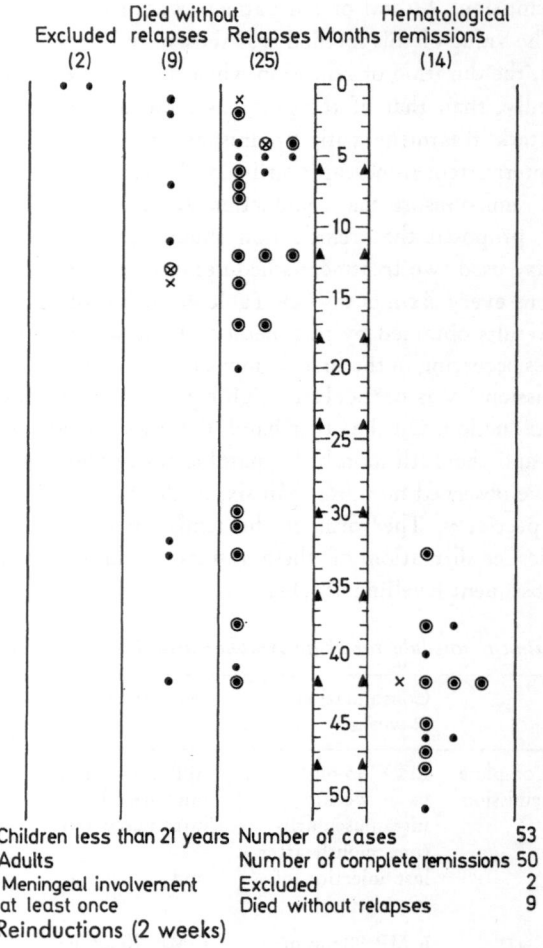

Fig. 55. Protocol 02 LA 64. Present condition of 52 patients treated (December 1968)

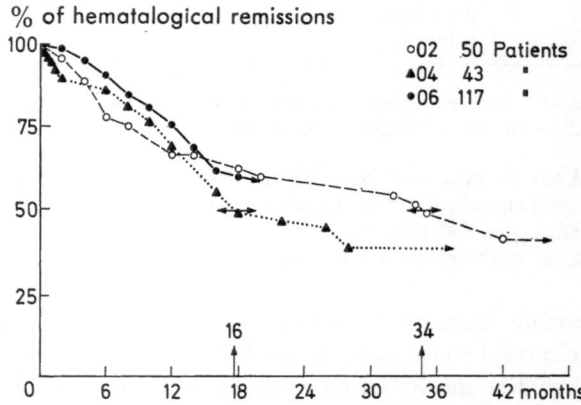

Fig. 56. Acute lymphoblastic leukemia. Calculation of the cumulative rate of blood remissions by actuarial methods (protocols 02 LA 64, 04 LA 65 and 06 LA 66). Duration of first remission in months

Fig. 57, only four relapses out of 17, or a little less than one third, occurred during the first seven months; on the other hand, there was still a drop in the curve from the 12th to the 18th month.

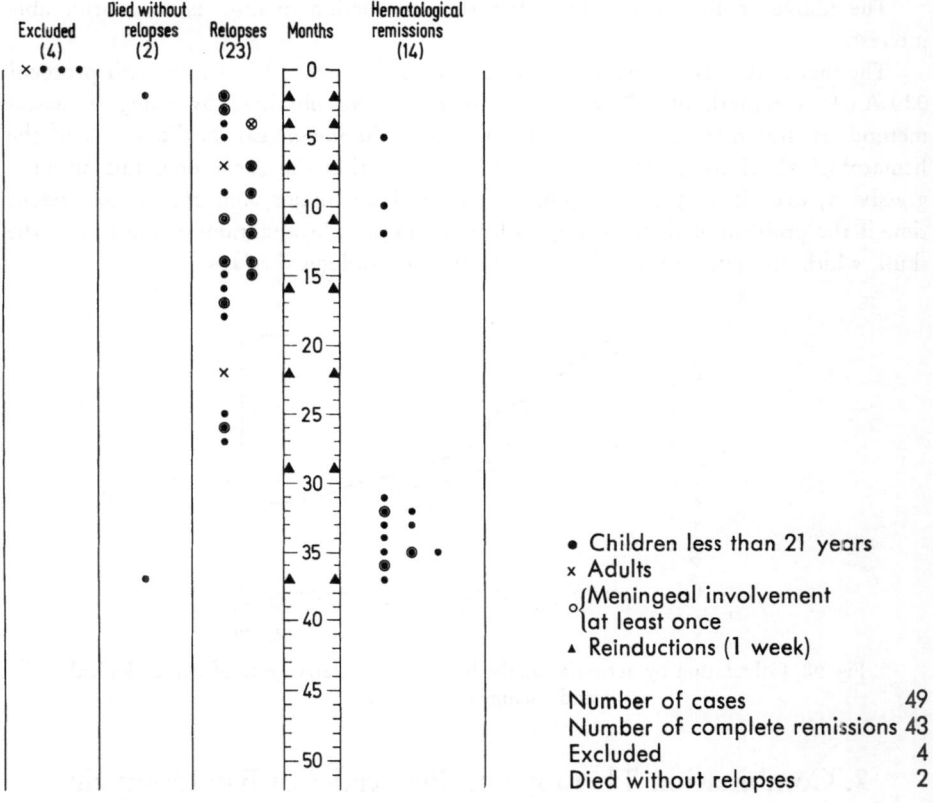

Fig. 57. Protocol 04 LA 65. Present condition of 45 patients treated (December 1968)

There are two possible explanations for this. It may be that the period over which reinduction was carried out in Protocol 04 was too short for a period of five or six months between reinductions. Alternatively, it may be that combination therapy with vincristine, prednisone, purinethol and methotrexate is incapable, however it is applied, of achieving more than 50% complete remissions of about 26 months.

With both protocols, regular examination of the cerebrospinal fluid showed meningeal involvement in many cases; this was always initially responsive to treatment with methotrexate, injected intraspinally. These meningeal involvements followed a variable course. At times, local relapse could be prevented by regular monthly intraspinal injections. At other times, the meningeal localization followed a subintrant course, leading to major changes in the fundus oculi and endangering the patients' sight. The treatment of such diffuse localizations, and in particular the place of radiation treatment of the central nervous system in such treatment, has still not been definitely established.

If actuarial methods are employed, the preliminary results can be expressed in two different ways. If the hematological relapses alone are considered, it is found that the median duration of complete remissions has still not been reached at 34 months for protocol 02 LA 64, or at more than 16 months for protocol 04 LA 65 [133].

The above results demonstrate that the reinduction method is of considerable interest.

The fact that only six relapses have occurred after the 30th month with protocol 02 LA 64 is remarkable. Nevertheless, if the curve obtained by using actuarial methods is drawn on the basis of defining both the meningeal localization and the hematological relapse as relapses, it will be found that the curve does fall off progressively, even if very slowly (Fig. 58). One of our major concerns at the present time is the problem of these meningeal localizations, or localizations at the base of the skull, which often occur several months before hematological relapse.

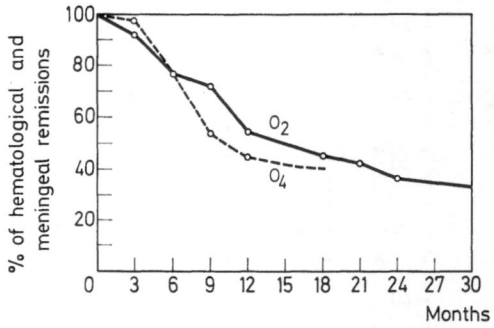

Fig. 58. Calculation by actuarial methods of the cumulative rate of hematological and meningeal remissions

2. Combination Therapy and Properties of Rubidomycin

In brief, the most important properties of rubidomycin are as follows:

1. rubidomycin can very quickly induce remissions in acute leukemia;

2. it can kill a very large number of leukemic cells;

3. it causes fundamental and serious changes in the bone marrow;

4. its administration over long periods often leads to serious cardiac complications.

As a result of the foregoing:

1. rubidomycin must be used mainly as an inducer drug, and especially when quick action is needed;

2. it can probably be used for reinduction purposes, provided that the number of reinductions is not too high;

3. it must never be used in maintenance treatment.

A number of different protocols based on these properties have been investigated. We shall consider here only those which have been used with an adequate number of patients.

3. Acute Lymphoblastic Leukemia

a) Use of Rubidomycin, Vincristine, and Prednisone in Combination

Rubidomycin was included in the protocol for systematic reinduction with vincristine and prednisone, mentioned above, in the hope both of increasing the already high proportion of complete remissions and of increasing the duration of the first remission [21]. The treatment used was as follows (protocol 06 LA 66).

b) Induction Treatment

Daily oral doses of prednisone (3 mg/kg or 100 mg/m²) were combined with weekly perfusions in which 60 mg/m² of rubidomycin and 1 mg/m² of vincristine during the first week, and 2 mg/m² during the following weeks, were injected successively into the tube used for an intravenous perfusion of physiological solution. The use of this drug combination was continued until complete remission was achieved.

c) Maintenance Treatment and Reinductions

Fifteen days after complete remission (less than 5⁰/o of marrow blast cells) had been achieved, a repeat perfusion with rubidomycin (30 mg/m²) and vincristine (2 mg/m²) was carried out, followed by maintenance treatment with a combination of 6-mercaptopurine (90 mg/m² orally per day) and methotrexate (15 mg/m² per week) by the intramuscular route (this latter drug was not introduced, however, until the reinduction treatment scheduled for the first month had been completed). This treatment was alternated with systematic courses of treatment lasting one week in which the drugs used for induction purposes were again administered [prednisone daily for 7 days, rubidomycin (30 mg/m²), and vincristine (1 mg/m² on the first day and 2 mg/m² on the seventh day)]. Reinduction treatment was carried out in the 1st, 2nd, 4th, 7th, 11th month, etc. From the first month onwards, this treatment was combined with the systematic intraspinal administration of methotrexate (two injections with each reinduction and a monthly injection between the reinductions). Methotrexate was not administered intramuscularly in the week in which it was intraspinally. The treatment schedule is summarized in Table 47.

During the remission period, a blood test was carried out at weekly intervals. If a leucopenia of less than 2000 per mm³ was found, or clinical signs such as mouth ulcers, high temperature, or signs of pulmonary infection, the intramuscular injection of methotrexate was postponed. A careful clinical examination, a determination of the bone marrow picture, and an examination of the cerebrospinal fluid were carried out at monthly intervals.

d) Patients Treated

For the sake of simplicity, the cases treated during the first stage of the development of the disease will alone be considered here. Complete remission was obtained by this treatment for relapses in patients resistant to vincristine and/or rubidomycin alone. As might have been expected, these remissions were of fairly short duration, but the cases treated were too diverse in character to enable us to study them properly.

Table 47. *Treatment schedule Protocol PARIS 06 LA 66*

Induction $\longrightarrow$ CR $\longrightarrow$ Maintenance		Reinductions [d]
Prednisone + Vincristine + Rubidomycin	i.v.	0.5^{a}—1—2—4—7^{c}—12—18—24—36 month then every 6 months:
Prednisone 100 mg/m² daily orally (3 mg/kg)	6-MP 90 mg/m² daily orally (2.5 mg/kg)	Prednisone 100 mg/m² daily for 7 days (14 days [c])
Vincristine 1 mg/m² 1st week then 2 mg/m² weekly	+ Methotrexate 15 mg/m² weekly (0.5 mg/kg intramuscularly)	Vincristine 1 mg/m² on 1st day 2 mg/m² on 7th day (and 14th [c])
Rubidomycin 60 mg/m² weekly (2 mg/kg)	+ Methotrexate 10 mg/m² spinally once monthly	Rubidomycin 30 mg/m² on 1st and 7th day (1 mg/kg) and 14th [c]
		2 spinal injections of methotrexate [b]

[a] One injection only of rubidomycin 30 mg/m² (1 mg/kg). Vincristine (2 mg/m²)
[b] Intraspinal methotrexate 8 and 10 mg/m² (0.2 and 0.3 mg/kg) after temporary arrest of maintenance treatment.
[c] Reinductions lasting 14 days from the 7th month on.
[d] After temporary arrest of maintenance treatment.

Thus 130 patients—39 adults and 91 children under 15 years of age—were treated, at the onset of the disease, according to the protocol Paris 06 LA 66. The sex distribution of the patients is shown in Table 48; as will be seen, there is a fairly marked predominance of the male sex.

Table 48. *Distribution by age and sex of 130 cases of acute lymphoblastic leukemia treated according to protocol 06 LA 66 (induction of complete remission by the triple combination of prednisone, vincristine, and rubidomycin)*

	Adults	Children [a]	Total
Male	33	54	87
Female	10	33	43
Total	43	87	130

[a] Children aged less than 15 years.

Our patients showed certain special clinical features; thus there were 11 patients with a tumoral form of the disease characterized by the large volume of the mediastinal adenopathies, the enlargement of the liver and spleen, and the large volume of the peripheral adenopathies. In four cases leukoblasts were found initially in the cerebrospinal fluid, while in the case of one child, fulminating hemiplegia was accompanied by an excess of protein in the cerebrospinal fluid only, which was corrected by means of intraspinal methotrexate and radiotherapy.

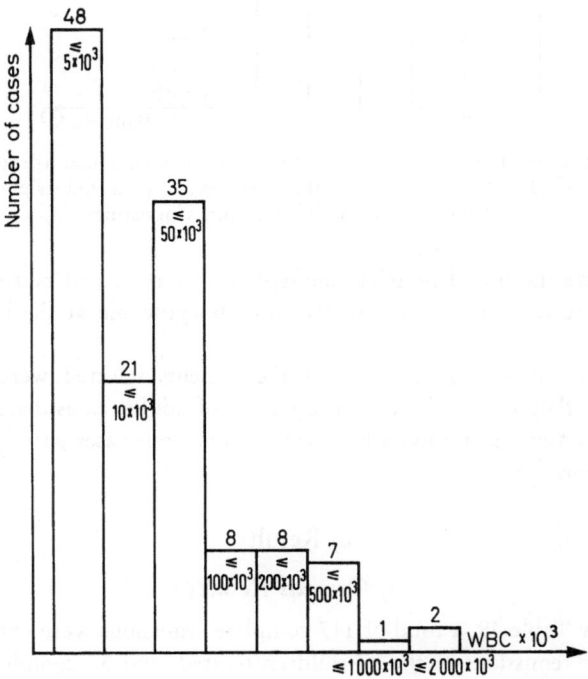

Fig. 59. Distribution of patients (130 cases) according to initial white cell count (protocol 06 LA 66). Acute lymphoblastic leukemia treated with the triple combination of rubidomycin, prednisone and vincristine

We shall not deal with the determination of the initial hemoglobin levels. Our patients had often been given transfusions before they were admitted to the department, and earlier work had shown that this factor was of no value in prognosis.

Fig. 59 shows the initial leukocyte counts for our patients. It will be seen that there were 32 cases in which hyperleukocytosis was present, the number of leukocytes being greater than 100 000/mm³, while in seven cases it was greater than 200 000/mm³.

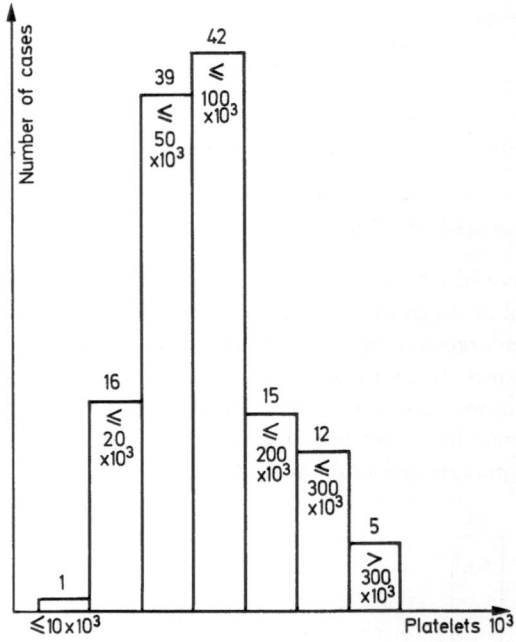

Fig. 60. Distribution of 130 cases according to initial platelet count (protocol 06 LA 66). Acute lymphoblastic leukemia treated in the first attack with the triple combination of rubidomycin, prednisone, and vincristine

Fig. 60 shows the initial platelet counts, from which it will be seen that most of the patients were severely or moderately thrombocytopenic at the beginning of the treatment.

In conclusion, it would appear that the patients selected were relatively unpromising, since they included a high proportion of adults, cases characterized by the presence of large tumors and hyperleukocytosis, and four cases with initial meningeal local involvement.

e) Results

α) Overall Results

As shown by Table 49, a total of 117 complete remissions were obtained, comprising 85 complete remissions in the 87 children treated, and 32 complete remissions in the 43 adults treated. It should also be pointed out that, of the seven failures, two patients died shortly after the first injection had been given, so that it was impossible to judge whether the treatment had had any effect. One of these patients died the day following the first injection, probably as the result of too violent a cellular lysis (350 000 white cells); the treatment with antileukemic drugs had not been preceded by the administration of allopurinol nor by forced diuresis. The second of these two

Table 49. *Results obtained in children and adults by the triple combination of prednisone, vincristine, and rubidomycin in treatment inducing complete remission (ALL)*

| | Adults | | Children | | Total |
	M	F	M	F	
Complete remission	22	10	52	33	117
Incomplete remission	—	—	—	—	—
Failure					
Rapidly fatal	2	—	2	—	13
Fatal aplasia	5	—	—	—	
Total resistance	4	—	—	—	
Total	33	10	54	33	130

patients died suddenly, the clinical picture being that of a meningeal hemorrhage, during his first week in hospital. In two cases, those of men of 50 and 75 years of age respectively, one died in a coma after the second injection, the other showing signs of septicemia associated with aplasia, after the fifth injection. In another case classified as a failure, a hematological remission had, in fact, been achieved by the 65th day, but the patient died nine days afterwards of an acute lung condition; his immune responses had been markedly depressed, as shown by the delayed-type hypersensitivity in his reactions and examination of the circulating antibody. One patient relapsed during the week following the remission. He was counted as a failure, as the subsequent application of the same protocol proved to be without effect. Only four of the 130 patients turned out to be completely resistant after this drug combination had been administered for six weeks.

β) Inducement of Complete Remission (117 Cases)

The treatment takes effect remarkably quickly. In general, the hemoblasts disappear from the marrow and complete remission is induced after a small number of perfusions of the combination of vincristine and rubidomycin, as follows:

1 perfusion was required in 30 cases
2 perfusions were required in 52 cases
3 perfusions in 23 cases
4 perfusions in 5 cases
5 perfusions in 6 cases; in one of these cases there was considerable enlargement of the liver and spleen; in another case, the initial leukocyte count was 510 000/mm^3
6 perfusions in 1 case.

In almost all the cases (82 out of 117), complete remission was preceded by marked leukopenia, often with a count of less than 1000, showing the existence of severe temporary hypoplasia, the average duration of which was four days.

Thanks to vigorous treatment with antibiotics, and in spite of the fact that high temperatures were observed in certain cases, this initial leukopenia was the cause of clinical complications in only four cases. In one adult patient, a leukopenia with a count of less than 500, which lasted for a week after the first perfusion, activated an abscess infected by *Klebsiella pneumoniae,* and this in turn caused the death of the patient one month later, while he was in hematological remission, a few days after the reinduction carried out on the 15th day.

In the case of another adult, peripheral pancytopenia and aplasia of the bone marrow appeared on the 10th day; the second injection of rubidomycin and vincristine had not been carried out, because the marrow was poor in cells and there were the first signs of dangerous leukopenia. Septicemia associated with Gram-negative bacteria and extremely serious in character (hyperthermal collapse) was controlled only by means of very vigorous resuscitation measures (transfusion of leukocytes, refrigeration, and massive doses of antibiotics).

In the case of two children, streptococcal septicemia was rapidly brought under control by vigorous treatment with antibiotics.

γ) *Duration of Remissions*

Fig. 61 shows the condition of these patients in December 1968.

Eleven patients have died, but not from any form of hematological relapse. We have already mentioned the first case, in which patient died of an abscess infected with *Klebsiella pneumoniae,* which developed during the initial stage of leukopenia, after the reinduction carried out on the 15th day.

Two other cases were those of children who died of severe lung conditions after the reinduction carried out in the 1st and 2nd months; with these children, each perfusion had been followed by a particularly marked leukopenia. Greater care should have been taken. Similar accidents had already occurred during consolidation treatment with 6-mercaptopurine and methotrexate, after reinduction with prednisone and vincristine. The precise character of these diffuse lung conditions has never been determined, in spite of the careful virological investigations that have been carried out. In particular, no cytomegalic inclusions have ever been found.

In one case, the patient died suddenly after the reinduction carried out in the first month. In general, the toxic effect of rubidomycin on the heart is seen only at higher doses than he was given, and the clinical signs are very different. Another mystery in the case of this patient was that hematological examinations gave normal results (there was, however, a meningeal relapse), while a histological examination of the bone marrow indicated that a relapse had occurred.

Two further cases were those of adults who died with signs of severe infection in the 2nd and 6th months of remission. No hematological relapse could be detected a few days beforehand, and as these patients died in the provinces no pathological examination was carried out and no precise conclusions can be drawn.

Eleven patients were eliminated from this study during the first six months of the course of the disease, either because maintenance treatment was not administered (often under the influence of a quack), or because of negligence on the part of the patient's family, or because the patient left the area and was lost track of.

Thirty-one patients relapsed between the 1st and 18th months.

At the present time, 64 patients are still in remission; their present condition is summarized in Fig. 61.

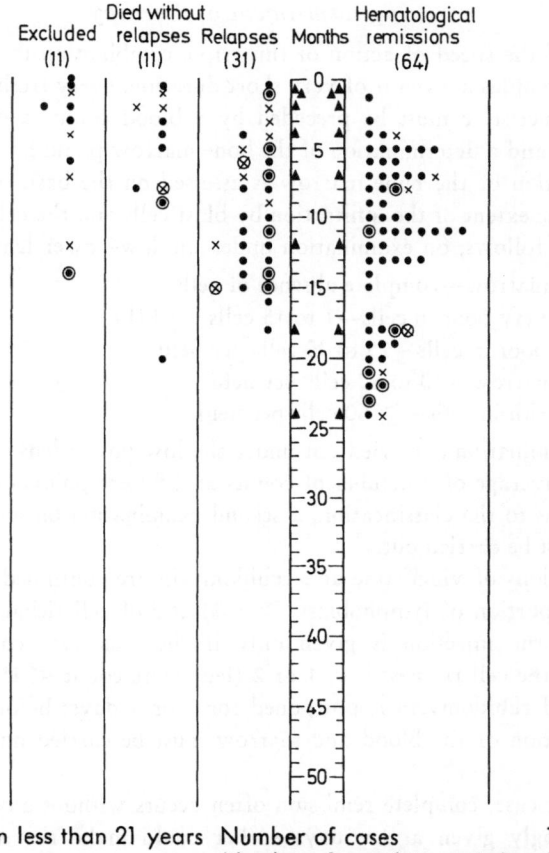

Fig. 61. Duration in months of complete remission. Protocol Paris 06 LA 66. Situation in December 1968

f) Tolerance

α) The Initial Phase

Clinically, the treatment was generally well tolerated, but temporary alopecia occurred with more than half the patients.

The perfusions were sometimes followed by attacks of fever. The speed with which leukopenia, often severe in character, occurred made hospitalization necessary, since treatment with antibiotics is the only way of preventing infections; only two of our patients died from this cause. In three cases with initial aplasia and a leukocyte count of less than 1000 when treatment was started, there was an almost immediate increase in the leukocyte count under treatment.

The fall in the platelet count seems to be less regular, less severe in character, and of shorter duration than the fall in the leukocyte count.

β) Adjustment of Therapy

In view of the speed of action of this triple combination therapy, and the possibility of severe aplasia, even if of very short duration, every fresh injection of rubidomycin and vincristine must be preceded by a blood count, a determination of the blood picture, and a determination of the bone marrow picture.

The condition of the bone marrow is assessed on the basis of the two following criteria, viz the extent of the infiltration by blast cells and the cell richness. The latter is classified as follows, on examination under the low-power lens of the microscope:

0 aplastic marrow—complete absence of cells
1 marrow very poor in cells—1 to 15 cells per field
2 marrow poor in cells—15 to 30 cells per field
3 normal marrow—30 to 60 cells per field
4 marrow rich in cells—$\geq$ 60 cells per field.

As the examination is carried out under the low-power lens, the values found are those for the average of a number of counts at different points on the smear. If there is any doubt as to the classification, a second examination on marrow from a different region must be carried out.

The perfusions of vincristine and rubidomycin are continued only if the marrow is blastic (proportion of lymphoblasts $> 5\%$) and of cell richness 3 or 4. If the cell richness is 2, the injection is given only if the leukocyte count is greater than 1500/mm³. If the cell richness is 0, 1 or 2 (leukocyte count < 1500), the injection of vincristine and rubidomycin is postponed for 2 or 3 days; before it can be given, a fresh examination of the blood and marrow must be carried out and the same precautions taken.

In this last case, complete remission often occurs without any fresh injection. An injection wrongly given a few days earlier could make the aplasia total and be responsible for a fatal septicemia.

If after two weeks of treatment the bone marrow remains at stage 3 or 4 before the third injection is given and contains more than 50% of lymphoblasts, the doses given in the third injection should be increased to 3 mg/m² of vincristine and 90 mg/m² of rubidomycin. These doses should be repeated in further injections if the fall in the number of blast cells is 65% or more of the initial figure. If the bone marrow stage is 3 or 4, the dose should be further increased to 4 mg/m² of vincristine and 120 mg/m² of rubidomycin.

At that stage it is a prudent measure to prescribe barbiturates and Diazepam to prevent any possible convulsive attacks brought on by the vincristine.

γ) Tolerance of Reinduction

All the patients but two of those observed tolerated the booster injection on the 15th day well. In one it may perhaps have aggravated the Klebsiella lung abscess he harbored; in the other it was followed by a virus pneumonia.

Reinductions were generally well tolerated. They caused no leukopenia; on the contrary, the number of leukocytes often rose between the first and the second perfusion.

Nevertheless, after the reinduction in the first month, when maintenance treatment had already begun, one patient had an acute lung condition preceded by jaundice that was probably due to the transfusion, a relapse of tuberculosis, and blood maturation disturbances (a transient increase in the myeloblasts and undifferentiated marrow elements) accompanying a severe otitis. During the reinduction of the second month we observed in one case a lung condition that was probably viral in origin, and in another already mentioned an acute lung condition preceded by aplasia and accompanied by monocytosis and marrow maturation disturbances that at first gave the impression of a relapse. In one case the amount of Purinethol administered was excessive (3.5 mg/kg instead of 2.5) and methotrexate had been given immediately after the booster injection of the 15th day—a practice that we gave up after the first mishaps. In all other cases the tolerance was perfect, making it possible for patients to receive their reinduction treatment while ambulant.

g) Conclusion

In nearly 90%/o of cases of acute lymphoblastic leukemia complete remission can be obtained by a combination of prednisone and vincristine. The remission takes place on the average between the third and fourth week of treatment.

The new combination including rubidomycin cannot be assessed on the percentage of remissions obtained (which is more than 90%/o). Remission, however, takes place more rapidly—between the first and third week—, and was easily obtained in very unfavorable forms of the disease such as the tumoral and the hyperleukocytic forms. Remission was obtained in 85 of 87 cases of children.

The value of the triple combination of prednisone, vincristine, and rubidomycin seems to be confirmed by the curve of the distribution of relapses in time and especially by the rarity of early relapses. It is difficult at present to judge whether among the patients treated there will be a fair number of the very long remissions that have been the subject of the rare but interesting observations recorded in the literature [27, 48, 49, 61].

h) Treatment of Relapses

The same schedule of treatment was applied to the treatment of acute lymphoblastic leukemia in relapse. We also added to protocol 06 LA 66 the cases of acute lymphoblastic leukemia in the first developing phase in which the patients, before being admitted to hospital, had been given more than three weeks' treatment with prednisone or several days with a combination of prednisone and vincristine but without rubidomycin. The results were again remarkable (Table 50) since, out of 69 cases treated, 59 complete remissions were obtained (85%/o). These patients had not had any rubidomycin previously, but in their relapses (43 cases, 37 complete remissions) had always had prednisone and vincristine as attack treatment and in reinductions (protocols 02 and 04) and 6-mercaptopurine and methotrexate for maintenance treatment.

On the other hand, this same protocol 06 LA 66 proved to be ineffective or of little effectiveness in the few relapses that occurred in patients who had already had the triple combination of prednisone, vincristine, and rubidomycin for inductions and reinductions. The duration of remission in these relapsing forms of the disease was generally much shorter.

Table 50. *Acute lymphoblastic leukemia treated by the triple combination of prednisone, vincristine, and rubidomycin: relapsed patients and patients returning to treatment (first attacks treated previously)*

Developing phase	Complete remission	Eliminated	Failure	Total
Initial (returning to treatment)	21	4	1	26
1st relapse	30		1	31
2nd relapse	3	1	3	7
3rd relapse	4			4
4th relapse			1	1
	59	5	6	69

"Eliminated" cases are those of patients treated at a very advanced stage in the disease and dying before the treatment could be effective.

i) Other Trials

A very similar schedule of treatment was used later by G. MATHÉ [172, 174] in the treatment of acute lymphoblastic leukemia and in the acute leukemic syndromes occurring during the development of lymphosarcoma; it consisted of 100 mg/m² of prednisone daily, 0.1 mg/kg of vincristine weekly, and 0.75 mg/kg of rubidomycin on the first two days of every week. The results he obtained in a so far limited number of cases were comparable with ours. MATHÉ et al. stress, as we do, the speed with which remission takes place and the general failure after a thorough search to find any leukemic cells after treatment with this triple combination. After prednisone alone or after prednisone and vincristine such cells are generally found. HOWARD [126] also used a combination of prednisone and rubidomycin.

k) Other Combinations Used in Treating Acute Lymphoblastic Leukemia

Other schedules of treatment are at present under study for first relapses in patients with acute lymphoblastic leukemia (protocols 02 LA 64, 04 LA 65, 06 LA 66) who are suspected of having become resistant to vincristine—when relapse occurs shortly after reinduction that includes vincristine—or who, because of serious and slowly regressive polyneuritis, cannot be given vincristine, a neurologically toxic drug when taken over a long time [141]. These schedules do not include rubidomycin except when the total dose does not exceed 30 mg/kg (protocols 02 and 02 bis 67).

Induction treatment consists of either cytosine arabinoside, 30 mg/m² daily given intravenously and rubidomycin, 90 mg/m² weekly, or cytosine arabinoside, 30 mg/m² daily given intravenously, and rubidomycin, 60 mg/m², given intravenously daily for three consecutive days and cytoxan, 500 mg/m², given intravenously on the first day. A second and then a third course of treatment of this kind are carried out every 7—10 days according to the state of the bone marrow and the extent to which the leukocyte count falls.

Maintenance treatment consists of either a combination of cytosine arabinoside (30 mg/m² weekly in subcutaneous doses) and methyl-GAG (350 mg/m² weekly intramuscularly) or cytoxan, 1000 mg/m² weekly intravenously or intramuscularly.

4. Acute Granulocytic Leukemia

a) Initial Combination of Rubidomycin with other Drugs

We have been led to combine rubidomycin with other active compounds because of the nature of its success in inducing complete remission in acute granulocytic leukemia: its success is partial—50% complete remission, but often fatal aplasia and difficulty in managing the drug. The interesting results obtained with the so-called "quadruple" form of treatment [230] induced us to combine the two methods.

The "quintuple" schedule of treatment under study at present is as follows:

cytosine arabinoside, 30 mg/m² three times weekly in a four-hour perfusion of isotonic glucose solution, alternating with:

methyl-GAG, 350 mg/m² three times a week intravenously

6-mercaptopurine, 90 mg/m² daily orally

prednisone, 40 mg/m² daily orally

rubidomycin, 60 mg/m² intravenously once weekly.

The number of patients we have treated by this method is still too small for us to report our results. The combination, however, seems to be satisfactorily tolerated.

Other schedules of treatment can be drawn up, as for example:

cytosine arabinoside, 100 mg/m² intravenously daily

rubidomycin, 60 mg/m² twice weekly intravenously.

b) Prolongation of Complete Remission

When complete remission is obtained, there are several possible courses of action:

1. As in systems of the VAMP type, all treatment could be stopped until a relapse occurred.

2. Treatment could be limited to periodic reinductions of the same type, consisting for example of a week's treatment every month with the same drugs as those used for the induction, with no treatment in the interval.

3. Maintenance treatment could be initiated, stopped periodically for reinductions.

The maintenance treatment so far chosen combines, as for acute lymphoblastic leukemia, 6-mercaptopurine and methotrexate. The reinductions are carried out with rubidomycin and methyl-GAG (page 119).

Other combinations of drugs are under study: for maintenance treatment, subcutaneous cytosine arabinoside, 30 mg/m² three times a week plus 6-mercaptopurine, 90 mg/m² daily, given orally; or cytosine arabinoside, 30 mg/m² weekly, given subcutaneously, plus methyl-GAG, 350 mg/m² weekly, given intramuscularly; and for reinductions, rubidomycin, 30 mg/m² on the first and seventh day, given intravenously, plus methotrexate, 15 mg/m² on the third and fifth day, given intramuscularly.

Chapter 11

Cytological and Immunological Study of Acute Leukemia Treated with Rubidomycin

1. Cytological Study

The various therapeutic methods applied to acute leukemia undoubtedly alter the leukemic cells as well as the normal cells. Marrow megaloblastosis was seen, as had been anticipated, as soon as antifolic agents were employed 20 years ago, and it has been known for a long time that the classification of a leukemic cell as a lymphoblast or myeloblast, often easy at the beginning, is totally impossible when treatment has been started [152]. On the whole, however, any attempts at analysis of these changes have been disappointing and have either ended up by being suspended or have been continued in a listless fashion.

On the other hand, considerable changes in the cell have been observed with rubidomycin in acute lymphoblastic leukemia. They are easily recognized and are useful in the conduct of the treatment, which is difficult.

a) Acute Lymphoblastic Leukemia [153, 154]

The morphological abnormalities (Fig. 62) concern especially the lymphoblasts and to a minor extent the normal cells of the bone marrow.

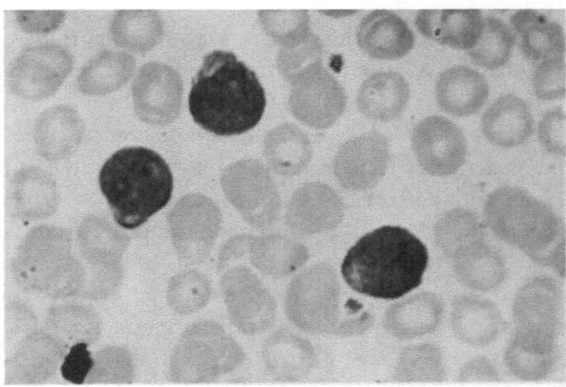

Fig. 62. Lymphoblast of acute leukemia before treatment (bone marrow)

The lymphoblasts are often greatly changed. The changes affect the nucleus and the cytoplasm and result in the formation of large cells (up to three times the normal size) that can be classified in two groups:

1. Lymphoblasts that are still recognizable as such. Large in size, with a fairly extensive cytoplasm, blue, without inclusions but sometimes with vacuoles, of irregular contour, their nucleus may still retain the typical dense chromatin of lymphoblasts as well as their very ill-defined nucleoli, or the chromatin may begin to expand, showing fine or thick meshes, well delimited, and several clearly visible pale nucleoli (Fig. 63).

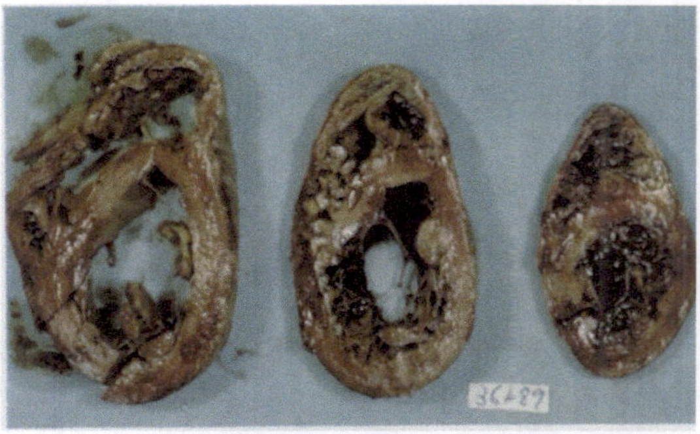

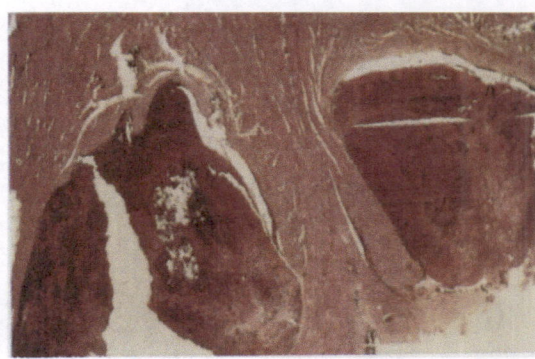

Fig. 43. Transverse section of the heart. Note thinning of the ventricular wall and the mural thrombus

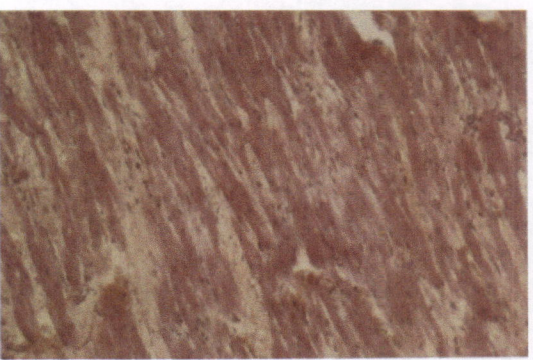

Fig. 44. Thrombosis partially obstructing the ventricular cavities (magnification ×10)

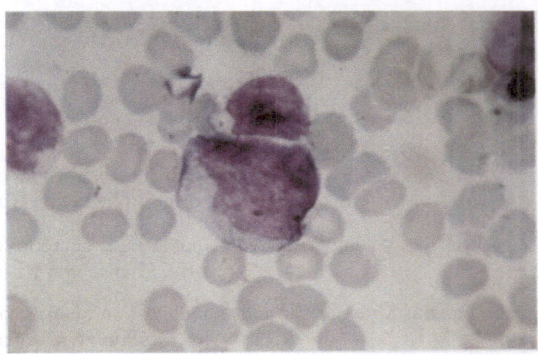

Fig. 47. Vascular degeneration of the myofibrils, myocarditis (magnification ×100)

Fig. 63. Lymphoblasts of large size still recognizable after treatment with rubidomycin for several days (bone marrow)

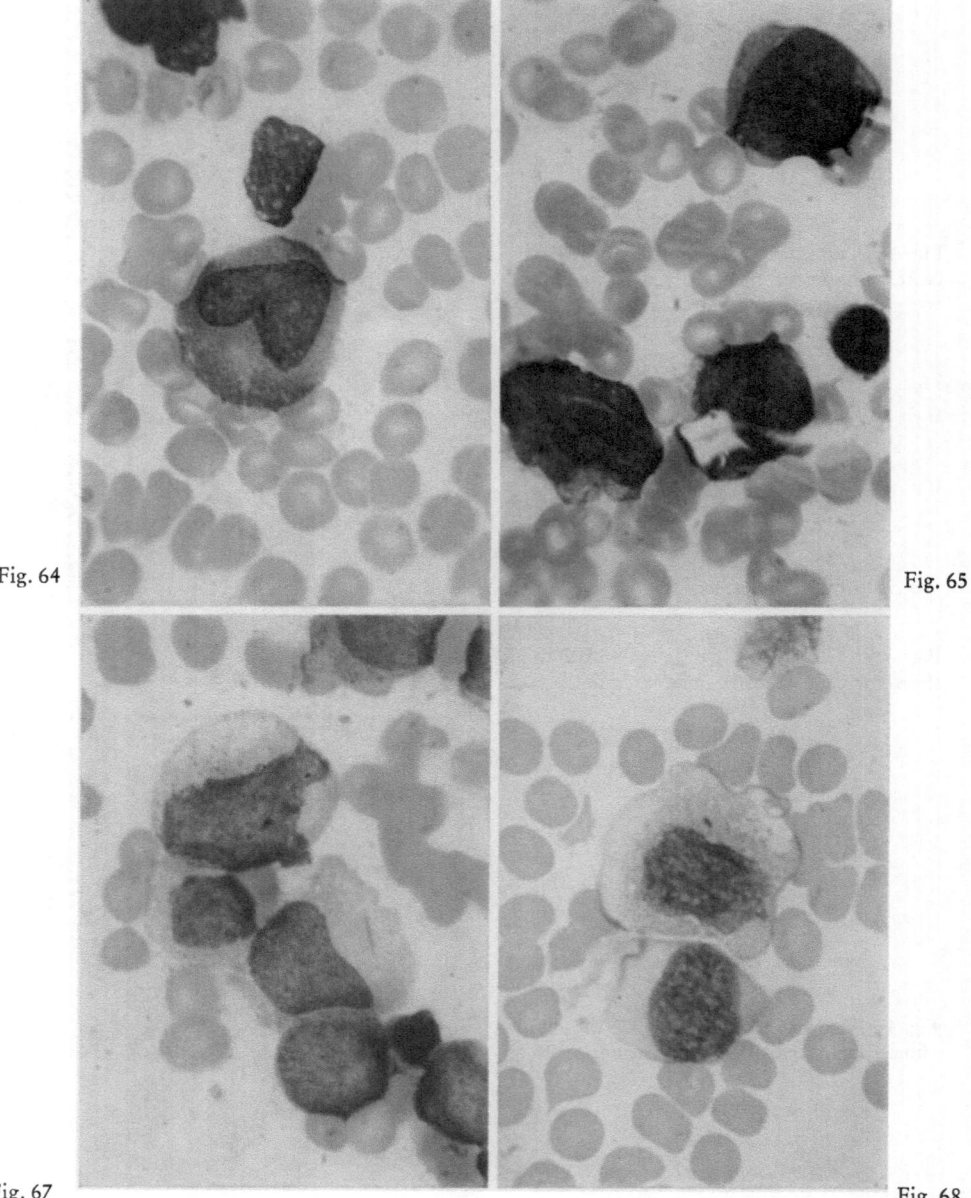

Fig. 64

Fig. 65

Fig. 67

Fig. 68

Fig. 64 and 65. Acute lymphoblastic leukemia treated with rubidomycin. Unclassifiable very deformed blast cells from the bone marrow

Fig. 67. Myeloblasts and hemoblasts in acute myeloblastic leukemia treated with rubidomycin. Blood elements with distinct changes

Fig. 68. Myeloblasts and hemoblasts in acute myeloblastic leukemia with frank changes caused by rubidomycin. Auer body still visible

2. Unclassifiable and very misshapen blast cells of very large size with an enormous nucleus of finely granular chromatin of regular texture and often one or two blue nucleoli (in fine, a nucleus of the "reticular" type) and a large amount of pale blue cytoplasm (Fig. 64 and 65).

All these elements are related to each other and appear to constitute a series of developing aspects of a relatively unmodified lymphoblast ending up as lysed cells leaving little but the base nuclei.

Normal marrow cells also show some abnormalities:

— In the erythrocyte series, some typical or intermediate megaloblasts may be found, but they are few in number and quickly disappear. Very large normoblasts are seen more frequently; their normally dense chromatin is very fragmented and they end up as red blood corpuscles containing Jolly bodies.

— In the granulocyte series, the elements are large and present the appearance observed at the beginning of or during megaloblastic anemia: large size, elongated nucleus with a loose chromatin network, and small granules in the cytoplasm. No overt abnormalities are observed in the megakaryocytes.

The chronology of these changes is interesting (Table 51).

The interval between the start of treatment and the changes in the blast cells is often short. In our first series it was less than five days in six out of ten observations.

The interval between the changes in the blast cells and the advent of aplasia is especially short: in nine out of ten cases it was less than three days: in only case was it long—15 days.

When aplasia occurs we have always seen it preceded by changes; aplasia seems to be a necessary stage in the progress towards complete remission. On the other hand, when the blast cells do not change or change very little, it seems that treatment is ineffective or hardly effective. There is no risk of sudden aplasia, but there is scarcely any hope of a remission.

If with an attack of leukemia there are any normal cells left, any minor abnormalities in them can be seen very early.

The duration of aplasia varies between one and two weeks.

The repair phase witnesses the return of the normal cell series all at the same time; the megakaryocytes appear just as soon as the red cell series and the granular series. This is the stage at which slight abnormalities can be seen in the granular series, but they disappear in less than a month. In the two cases classified as incomplete remission the aplastic phase was actually followed immediately by a relapse characterized by the presence of a scanty population of blast cells in a very impoverished bone marrow.

When a relapse follows a complete remission of a certain duration, the lymphoblasts are of fairly large size but easily recognizable.

These facts are to some extent suggestive; it would seem that when the treatment is going to have some effect its action is in most cases quick. But the most interesting changes are those of the blast cells; so far we have never seen such giant cells with other forms of treatment such as prednisone or vincristine. When they become observable they seem to herald the early advent of aplasia—an indication of obvious importance in the management of treatment.

Table 51. *Cell development in the marrow in 10 cases of acute lymphoblastic leukemia treated with rubidomycin*

	1	2	3	4	5 Complete remission	6 Complete remission	7 Complete remission	8 Incomplete remission	9 Incomplete remission	10 Partial failure
Interval between:										
Beginning of treatment and change in blast cells	4 days		13 days	3 days	4 days	4 days	4 days	7 days	5 days	1 day
Change in blast cells and aplasia	0 days		3 days	0 days	2 days	0 days	3 days	3 days	15 days	2 days
Start of treatment and hypoplasia or aplasia	4 days	4 days	16 days	3 days	6 days	4 days	7 days	10 days	20 days	1 day
Aplasia and repair stage					14 days	16 days	12 days	10 days	15 days	

b) Acute Granulocytic Leukemia [154]

We also studied cell development in acute myeloblastic leukemia treated with rubidomycin. The study was particularly thorough and covered eight cases. Of the eight patients, four had been treated for a first attack of the disease, four for relapses (two for a first relapse, one for a second and one for a third).

The morphological changes in the myeloblasts observed were slight or frank:

— They were slight in two cases, the myeloblasts being hardly changed and easily recognizable. Sometimes the nucleus became notched and monocyte-like in appearance, but this is a very frequent occurrence, often even before any treatment is instituted.

— In six cases the changes were frank, the myeloblasts being of large size. These large myeloblasts show two appearances:

In one the cells were very large and the ratio between the nucleus and the cytoplasm was high. The nucleus was bulky, very variable in shape, and often very notched, with fairly dense chromatin. The nucleoli remained numerous and clearly visible. The cytoplasm was very blue and the azure granules were clearly distinguishible.

In the other the cells were very large but the ratio between the nucleus and the cytoplasm was less marked. The nucleus was large, geometrical in shape, and often attached to the cell wall, and it had a fine, dense chromatin. The nucleoli were pale in colour and difficult to see. The cytoplasm was abundant, with wavy outlines, and blue-grey in colour; the granules contained in it were numerous but fine. These cells recalled the classic description of the leukemic "monoblast", and they could readily have been identified as such but for their initial appearance and the opportune appearance of Auer's bodies.

All the cells showing these changes are found close one to another and to cells that are little changed or are already completely lysed.

The normal marrow cells that may persist present the same abnormalities as in lymphoblastic leukemia.

The chronological development of these abnormalities shows characteristics rather different from those we observed with acute lymphoblastic leukemia (Table 52).

— In the six patients in whom frank changes were observed, the interval between the start of treatment and the appearance of the changes was fairly short: 4—8 days.

— But the changes in the blast cells did not inevitably herald aplasia:

Case 1: the changes began after five days of treatment but there was no aplasia and no remission.

Case 5: the changes began seven days after treatment but there was no aplasia and no remission.

Case 3: there were frank changes in five days, the blast cells disappeared gradually, and complete remission set in after 15 days with no aplastic phase.

— When aplasia sets in (Cases 6—8) after the blast cells change, it occurs after a somewhat longer interval (6—8 days).

— It should be particularly noted that the these changes do not represent an essential stage in the preparation of the complete remission. In Cases 2 and 7 there were only slight changes, but after 22 and 14 days respectively aplasia set in, preceding complete remission.

Table 52. Cell development in the marrow in eight cases of acute myeloblastic leukemia

	1	2	3	4	5	6	7	8
Previous treatment	CA+GAG →CR 6-MP+Pred.	CA+GAG →CR 6-MP+Pred.	1st attack	1st attack	1st attack	CA+GAG →CR 6-MP+Pred.	CA+GAG →CR 6-MP+Pred.	1st attack
Interval between start of treatment and change in blast cells	5 days overt changes	slight changes	5 days overt changes	4 days overt changes	7 days overt changes	4 days overt changes	slight changes	8 days overt changes
Interval between changes in blast cells and aplasia			gradual disappearance of blast cells	8 days		6 days	7 days	7 days
Interval between start of treatment and aplasia		22 days	no aplasia	12 days		10 days	14 days	15 days
Interval between aplasia and repair stage		10 days		15 days		15 days	8 days	9 days
Results	Failure	CR	CR	CR	Failure	CR	CR	CR

CA: Cytosine Arabinoside — GAG: Methyl glyoxal-bis-guanyl hydrazone — Pred.: Prednisone — CR: complete remission

— The duration of the aplasia was between 8 and 15 days.

— With remission there were often some abnormalities of the erythrocyte and granular series.

A final observation is that abnormalities of this kind—giant cells, histiocyte appearance—have already been clearly demonstrated with other drugs, as for example cytosine arabinoside.

These findings need comments rather different from those we made on lympho-blastic leukemia. The abnormalities observed are not new. They were inconstant, and their absence did not suggest that the treatment was ineffective, since complete remission could occur without any great changes in the myeloblasts. If changes did occur they did not necessarily indicate that aplasia was imminent; aplasia could fail to appear, either because the treatment had met with complete failure or because the disease was progressing favorably with a gradual disappearance of blast cells and a simultaneous reappearance of the normal elements of the bone marrow. When aplasia occurred, it was a little later than when the leukemia was lymphoblastic.

We might state as a conclusion to this preliminary study that it seems to us that the abnormalities in the lymphoblasts on treatment are interesting because they are peculiar to the disease and occur a short time before the aplasia, which appears to be a stage essential to the occurrence of complete remission. On the other hand, the abnormalities in the myeloblasts on treatment have no particular interest for the clinician.

2. Immunodepressive Power of Rubidomycin

Because it has an important secondary effect, the power of antineoplastic drugs to inhibit immune reactions deserves at present to be ranked among their characteristics. An immunodepressive action is in fact capable of increasing the risk of a complicating infection, and if the hypothesis of an immune reaction of the host against his tumor is justified, it could even be a factor that could in the long run affect the anticancer action adversely.

In the absence of any drug that is strictly specific for neoplastic cells, it can be said that there is at present hardly any cytostatic drug that is not also able to change the immune response.

This immunodepressive power varies greatly from drug to drug, the type of immune reaction called forth probably depending on the mechanism of action of the drug concerned. The immune reaction may be affected by interference with the bio-synthesis of DNA and RNA or by a change in protein synthesis—either by inhibiting the division of the immunologically competent cell or by hampering its differentiation by inhibiting antibody synthesis. An immunodepressive effect may also result in-directly from the toxicity of the drug to the cells taking part in the phagocytosis of antigen or in the inflammatory reaction.

Rubidomycin is an antimitotic drug that interferes with the metabolic functions of DNA and RNA [51, 115, 144, 148, 165, 190, 202] and is toxic to stem cells of the granular series [136, 139]. It can therefore justifiably be suspected of having a depressive effect on immune reactions. Studies on man and on animals have confirmed that this is so.

a) Studies in Animals

The work of MARAL et al. [165] provides some information on the effect of rubidomycin on certain immune responses in the mouse.

When rubidomycin is administered intraperitoneally in a dose of 1.25 mg/kg to R III mice 48 hours before and one and 48 hours after an injection of sheep red cells, it reduces considerably the level of hemagglutinating antibodies produced 10 days afterwards. This inhibition of circulating antibody production appears to be greater than that produced in the same mice by amethopterin and azothioprine.

Similarly, when AKR/RhO mice grafted with H2-incompatible (C57Bl/RhO) skin were given a daily intraperitoneal dose of 0.625 mg/kg of rubidomycin, it delayed to a slight extent (2—3 days) the rejection of the grafts.

The effect of rubidomycin on the *in vitro* transformation of lymphocytes has also been studied in dog lymphocytes in a mixed culture [165]. Inhibition of blast cell transformation was detectable at a concentration of 0.01 µg/ml, which produced no morphological cell lesion. The probably larger dose that would inhibit transformation induced by phytohemagglutinin is not known. This effect of rubidomycin is perhaps comparable with that of other antibiotics such as mitomycin C.

The study of the immunodepressive effects of rubidomycin in animals is undoubtedly only beginning. It deserves to be extended to animals other than mice and to other immune responses, particularly delayed hypersensitivity to other antigens. By comparing it with the effects of other drugs a clear idea could be formed of the immunodepressive effects of rubidomycin.

b) Studies in Man

From our own studies we have been able to assess the immune status of patients with acute leukemia on various courses of treatment [149] and especially on treatment with rubidomycin.

The results are still incomplete, but they nevertheless give some specific indications of the effect of rubidomycin on the immune responses.

The patients were treated with rubidomycin alone, in a daily dose of 1—2 mg/kg given intravenously, over a period of 3—8 days [136].

α) Effect on the Production of Circulating Antibody

The production of humoral antibodies was studied after the subcutaneous injection of 1 ml of Institut Pasteur poliomyelitis vaccine containing attenuated and killed type I, II, and III viruses. The serum antibodies against the three types of virus were determined separately before the injection and then 15 days after it by study of their ability to neutralize the cytopathogenic effect on HeLa or KB culture cells. In this technique a rise in the antibody level is regarded as significant if the antibody titer determined 15 days after injection of the vaccine is at least four times the initial titer.

Because the subjects had often had vaccinations previously, most of the antibody reactions studied were of the secondary type.

In a control series of 26 normal adult subjects, 24 (92%) showed a significant rise in at least one of the three types of virus antibodies after injection of the vaccine.

Six patients with acute myeloblastic leukemia treated with rubidomycin received an injection of poliomyelitis vaccine on the same day as or within 48 hours of the start of treatment. Of these six, four showed no detectable antibody response on the 15th day. All these patients, it should be observed, underwent a period of leukopenia. By way of comparison, out of 11 patients with acute lymphoblastic leukemia treated according to protocol 06 LA 66 [21] with a combination of prednisone and vincristine and, in more widely spaced doses, rubidomycin (2 mg/kg weekly), only four patients showed no antibody response.

In spite of the small number of cases studied, it can be seen that rubidomycin appears to be able to inhibit the antibody response when it is administered in daily doses. The slightness of the inhibiting effect in another course of treatment in which the doses were more widely spaced suggests that the effect on the antibody response is greatly reduced when the drug is administered with a longer interval between the doses.

β) Effect on Delayed Hypersensitivity Reactions

Previously acquired delayed hypersensitivity was tested by the intradermal injection of 0.1 ml of the following antigens: candidine diluted to 1 : 1000 (Institut Pasteur), IP48 tuberculin in a dose of 10 units (Institut Pasteur), streptokinase-streptodornase (Varidase, Lederle Laboratories) in a dose of 20 units of streptokinase, and the mumps control antigen (Institut Pasteur), which is a non-virulent extract of egg embryo.

A negative control was furnished by an injection of 0.1 ml saline or of histamine.

The skin reactions were read at 30 minutes, and then at 24, 48 and, in certain cases, 72 hours after the injection, three parameters being measured: the erythematous reaction, the amount of induration as assessed by palpation, and the presence or absence of edema. Here only the qualitative results are given, a positive reaction being defined as the presence of an induration of 3 mm or more after 48 hours.

The interpretation of reactions is sometimes hampered by ecchymoses in leukemic patients, who are often thrombopenic. Out of the 617 intradermal reactions performed, 7% were ecchymotic and could not be interpreted: the patients were omitted from the study except when they had ecchymoses centering on an inflammatory reaction.

The intradermal reaction to these antigens was investigated in the patients before they received any treatment, and then repeated at intervals during the course of treatment.

Before treatment, in a control series of 50 acute leukemias, of which 25 were acute lymphoblastic leukemia, 96% of the patients reacted to at least one of the four antigens and most reacted to two.

During induction treatment with rubidomycin alone or in combination with prednisone and vincristine [21], we saw no case of complete failure of reaction to these four antigens in 21 patients tested outside the periods of post-therapeutic aplasia.

On the other hand, we observed cases in which delayed hypersensitivity was abolished during aplasia induced by rubidomycin. Twenty-four patients were studied who had gone into a phase of aplasia as a result of rubidomycin used alone in a dose of 1—2 mg/kg daily. Aplasia was defined as a white cell count of 400 or less per mm³ and a bone marrow without cells. In three patients the presence of an ecchymosis prevented the skin reactions from being read. The results obtained with the other 21 patients (13 with acute myeloblastic and 3 with acute lymphoblastic leukemia, 3 with acute leukemia with promyelocytes and 2 with unclassifiable acute leukemia) are shown in detail in Table 53. Fifteen of the patients reacted to at least one antigen.

Table 53. *Delayed hypersensitivity reactions in acute leukemia during aplasia induced by rubidomycin*

Name	Age	Type of acute leukemia	Rubidomycin		DHS (4 Ag)	Course (interval in days)
			Total dose (mg/kg)	Duration (days)		
Dup	56	M	13	9	−	DCD + 6
Roy	67	M	8	5	−	DCD + 2
Nog	49	M	1	1	− ✕	DCD + 19
Leb	32	PM	11	12	− ✕	DCD + 2
Bla	25	PM	12	6	− ✕	DCD + 2
Sma	28	PM	4	4	−	DCD + 10
Riv	21	L	13	10	+	DCD + 10
Gom	21	M	44	65	+	DCD + 5
Bar	31	M	18	12	+ ✕	R° + 5
Sim	12	M	12	30	+ ✕	R° + 26
Rou	10	M	18	17	+ ✕	R° + 4
Gau	19	M	19	25	+ ✕	R° + 6
Lec	25	M	15	35	+ ✕	R° + 2
Pou	63	M	6	5	+ ✕	R° + 2
Leg	73	M	5	5	+	R° + 2
Des	47	M	18	23	+	R° + 10
Gil	12	Uncl.	10	12	+ ✕	R° + 3
Jou	23	Uncl.	8	5	+ ✕	R° + 7
Vig	9	L	32	80	+	R° + 2
Seb	15	L	3	7	+ ✕	R° + 19
Mar	13	M	3	3	+	R° + 6
50 controls	38.7	All types	Before treatment		96%+	

✕ Delayed hypersensitivity reactions (DHS) positive before treatment
R° Reversible aplasia. DCD Death

On the other hand, six (three with acute leukemia with promyelocytes, three with myeloblastic leukemia) did not react to any of the four antigens. The abolition of skin reactivity in the case of three of them was confirmed by the same intradermal tests as had been carried out before treatment and found positive; and it was probable in the case of the other three, given the high percentage (96%) of delayed hypersensitivity reactions observed before treatment in patients with acute leukemia of all types at an equivalent age.

Rubidomycin thus seems to be able in certain cases to abolish pre-established delayed hypersensitivity skin reactions that are considered to be particularly resistant to the action of most cytostatic drugs.

In the six patients with a negative delayed hypersensitivity reaction the aplasia was irreversible and caused their death in between two and 19 days. Of the 15 patients in whom the delayed hypersensitivity reaction was positive, in 13 the white cell count rose steadily within an average period of seven days after the tests and only two died from aplasia after five and ten days respectively.

The total dose of rubidomycin that converted the delayed hypersensitivity reactions to negative in six patients varied greatly from one patient to another, from 1 to 13 mg/kg, distributed over 1—12 days. This dose was no larger than that which induced aplasia without loss of the delayed hypersensitivity reaction in the other 15 patients and probably depends on individual variations in susceptibility to the drug.

These results suggest a relationship with the seriousness of the aplasia induced by rubidomycin. The abolition of the delayed hypersensitivity reaction does not seem to behave as a selective depressor of this type of immune reaction, which seems to be abolished only by toxic doses of the drug.

There is thus an obvious advantage in carrying out these intradermal tests on patients undergoing treatment with rubidomycin. Within 48 hours they make it possible to make a prognosis on the development of the aplasia that occurs after treatment. If the aplasia is accompanied by negative delayed hypersensitivity reactions, a transfusion of white blood cells is urgently indicated in view of the seriousness of the prognosis. In favor of the transfusion is the depression of cell immunological reactions, which improves its chances of being tolerated.

Rubidomycin appears therefore to have immunodepressive effects. In man it may inhibit the secondary antibody responses in doses approximating to those that induce aplasia. Only in toxic and even lethal doses does it abolish pre-established delayed hypersensitivity reactions.

Although knowledge of this effect of rubidomycin is still incomplete, the drug is clearly capable of inhibiting immunological reactions both in the mouse and in man. *A priori*, however, at least in man, it does not seem to be an immunodepressive drug that can be used as such, since the immunodepressive doses are too close to the toxic doses.

Chapter 12

Treatment of Chronic Leukemia, Sarcoma, Hodgkin's Disease, and Cancer with Rubidomycin

1. Chronic Leukemia

a) Chronic Myeloid Leukemia

Chronic myelocytic period. Many therapeutic agents, among which Busulfan and radiotherapy occupy the first rank but which also include 6-mercaptopurine, dibromomannitol, and hydroxyurea, can bring about good remissions in chronic myeloid leukemia. There are thus two important questions to answer in relation to the use of rubidomycin as a therapeutic agent: 1. can it bring about remission in this condition? and 2. what advantages does it have over the drugs already known? It is easy at present to answer the first of these questions, for rubidomycin is indeed very effective. It is difficult to answer the second.

Dosage. The doses employed varied from 1 mg/kg daily every two or three days to 2 mg/kg every other day or even every day, with a total dose of 6—12 mg/kg (median 10 mg/kg, average 9.4 mg/kg) given in 5—28 days (median 9 days, average 12 days).

Course of the disease. Frequency of remission. Good remissions were very frequently obtained [216]. The white cell count returned to normal, the myelemia disappeared, the splenomegaly was considerably reduced, the spleen sometimes appearing to become normal, sometimes remaining moderately enlarged; and the red blood cells returned to normal, though sometimes a moderate anemia persisted for some days after the cessation of treatment. Where platelet abnormalities (thrombopenia or thrombocytosis) existed, they could—especially in the case of thrombocytosis—be corrected. In three cases out of four with a heavy initial thrombocytosis, we have observed the platelet count return to normal. Successive bone marrow determinations showed a steady improvement and a marked fall in the level of marrow granulocytes. With rubidomycin, as with the other agents used for treating chronic myeloid leukemia, the Philadelphia chromosome was observed to persist during remissions. These results were obtained both for chronic myeloid leukemia treated during the first developing phase and for the disease treated at the time of a myelocytic relapse. Table 54 shows the results obtained with our 11 patients who had not been treated previously.

Special features of the remission. The remission was remarkable because of the following features: a) it occurred rapidly, between five and 10 days after the cessation of treatment, i. e., between two and three weeks after the beginning of treatment; this rapidity of action is comparable to that of Busulfan [217]; b) it seems to be of short

Table 54. *Effect of a course of rubidomycin in 11 cases of chronic myeloid leukemia not previously treated*

Name	Age and sex	Before treatment		Treatment		Results	
		WBC (10³/mm³)	Spleno-megaly	Duration	Total dose (mg/kg)	WBC (10³/mm³)	Spleno-megaly
Pod ...	33 M	288	+++	20 days	12 mg	1.8	0
Tes ...	53 M	82	++	8 days	10 mg	5.4	0
Dau ...	34 F	247	+++	10 days	9 mg	2.4	0
Pep ...	26 M	122	++	9 days	10 mg	5.4	0
Sta ...	14 M	460	Huge	28 days	12 mg	5	++
Leg ...	52 F	212	+++	10 days	9 mg	5.6	0
Pau ...	51 M	227	+	7 days	8 mg	4	0
Thi ...	28 M	110	++++	18 days	10 mg	2.8	++
Bri ...	56 M	243	Huge	9 days	10 mg	3.4	+
Fei ...	22 M	320	Huge	5 days	10 mg	3.4	++
Pet ...	42 M	184	++	6 days	6 mg	3.2	0

Splenomegaly: + palpable 1—3 cm below the costal margin; ++ 3—5 cm below the costal margin; +++ reaching the umbilicus; ++++ crossing the umbilicus. Huge: largely palpable in the right iliac fossa.

duration—of the order of one month—very probably much shorter than that induced by Busulfan; c) there seems to be no cross-resistance between rubidomycin and
the drugs usually employed in treating chronic myeloid leukemia.

Remission can also be obtained in cases refractory to other drugs. Rubidomycin
was administered to five patients in a myelocytic relapse. Three, whose disease had
been going on for 1 1/2 to 4 years, had never had rubidomycin, the other two had had
a course of treatment with it and had relapsed. In four cases very satisfactory results
were obtained: three complete remissions, one incomplete remission. Two of the
remissions were obtained in patients in a very poor general condition, suffering
from a high temperature, sweating, loss of weight, and thrombocytosis. The signs
indeed suggested the imminence of an acute transformation, and the condition had
been refractory to Busulfan and to rufocromomycin.

Indications for rubidomycin. With experience of rubidomycin still limited in the
treatment of chronic myeloid leukemia, it is difficult to set out the indications for its
use at present. The following comments may however be made [215].

Because of cardiac complications and of the danger of prolonged treatment with
rubidomycin, there does not seem to be ground for recommending it as the basic treatment and especially not the maintenance treatment for chronic myeloid leukemia.

But because it acts rapidly, it should be indicated in just those cases where rapid
action is needed, for example for patients with priapism or with an infarct of the
spleen.

Tolerance. With the method of administration we employed, we observed no
development of profound marrow hypoplasia. The risk of this occurring is certainly
not negligible, especially as rubidomycin has an undoubted delayed action.

In spite of the enormous mass of cells lysed, neither the blood urea or uric acid
nor the urinary uric acid reached very high levels, undoubtedly because abundant
liquid and Allopurinol were prescribed. In two cases digestive disturbances with
anorexia followed the injections. In three cases there was a partial loss of hair. And
a patient aged 62 years of age in an acute transformation phase was found to have a
myocardial infarction; rubidomycin injections had been continued to maintain the
improvement she had shown but had recently been stopped because of the appearance
of tachycardia. It is possible that rubidomycin (19 mg/kg injected in three months)
was responsible, although all clinical manifestations of cardiac insufficiency completely disappeared in two weeks. Three months after the rubidomycin was stopped
this patient, now on hydroxyurea for the past three months, is in an excellent clinical
condition and her blood picture is also excellent.

Acute transformation. (Blast cell crisis.) The acute transformation (blast cell
crisis) in chronic myeloid leukemia remains one of the most serious occurrences in
hematology. Although, as one of us pointed out as early as 1952, remission can be
induced with various therapeutic agents such as heavy doses of cortisone and also with
6-mercaptopurine, and hydroxyurea, the remission is always short and rapidly followed by either a fatal aplasia or the inexorable return of the blast cell crisis [14]. The
trials with rubidomycin were not very satisfactory. The dose employed was 2 mg/kg
twice weekly. In one case complete remission was obtained after administration of a
total dose of 8 mg/kg over two weeks; a white cell count of some 800 per mm^3 was observed but it lasted for a few days only. In another case the blood and marrow picture, the enlarged spleen, and the general condition were favorably influenced after a

total dose of 19 mg/kg in three months. In one case rubidomycin, prescribed in the middle of a terminal blast cell crisis, was ineffective.

The remissions obtained were comparable with those obtained with the other forms of treatment; they were short in duration and rapidly followed by a relapse and progression of the disease. It is possible that somewhat better results could be obtained by weaker doses and a wider spread of the doses over time. The most important step of all would be to intervene at an earlier stage. We are continuing to study ways and means of improving the treatment.

b) Chronic Lymphoid Leukemia

Chronic lymphoid leukemia does not seem to respond to rubidomycin. The therapeutic trials have, however, been restricted to serious cases—either refractory to other forms of treatment or characterized by hyperleukocytosis. Some of the latter cases had not had any treatment, and the failure of rubidomycin seems to indicate its ineffectiveness in chronic lymphoid leukemia. Perhaps additional trials would be worth carrying out.

2. Sarcoma of the Hemopoietic Organs

Rubidomycin is able to act beneficially on sarcoma of the hemopoietic organs, or at least on some of the sarcomas. Its effect on them seems to be less marked, less frequent, and less often complete than on acute leukemia. The number of cases treated, however, was not very great, and comparisons with the effects obtained by the therapeutic methods available at present have not been systematically carried out. Since this sums up the present position, it is difficult at present to determine the exact place rubidomycin may possibly have in the treatment of sarcoma of the hemopoietic organs.

a) Lymphoblastosarcoma

Rubidomycin has been used alone or in combination with other drugs.

α) Rubidomycin Used Alone

Rubidomycin was given in doses of 0.5 mg/kg daily for 10—20 days or of 1 mg/kg taken every other day, according to hemopoietic tolerance—with spacing out of the injections when the white cell count was below 3000 per mm³ and temporary cessation of treatment when it was below 2000 per mm³.

The positive data are rather scanty. J. S. BOURDIN [37], however, reported two remissions of short duration out of three cases treated. The complete failure reported by that author is somewhat surprising, since with him rubidomycin not only had no effect on lymph-gland tumoral formation and dermoepidermal nodules but also failed to prevent a marrow proliferation of blast cells at the same time as treatment was being given—proliferation that was later overcome by treatment with a combination of prednisone and vincristine. C. TAN and L. MURPHY [214] and J. CHAUVERGNE et al. [55] also report some cases of lymphosarcoma that reacted favorably but transiently to treatment with rubidomycin employed alone.

β) Rubidomycin Used in Combination with other Drugs

The drug treatment of the generalized forms of lymphosarcoma has not been strictly rationalized and the results—at least the long-term results— remain rather mediocre. Attempts are mostly restricted to arresting the developing phases of the disease by alkylating agents or combinations of corticosteroids and alkylating agents. No provision is made for consolidating the gains or maintaining the patient.

Encouraged by some successes obtained by the combination of prednisone and vincristine in equivocal forms of the disease in which the invasion of the bone marrow was so great that it suggested the diagnosis of acute leukemia, we decided to try to treat lymphosarcoma according to a schedule similar to that for acute lymphoblastic leukemia. We treated 15 patients according to protocol 06 LA 66, the same as that used for acute lymphoblastic leukemia [229]. The treatment consisted of oral prednisone in a daily dose of 100 mg/m² and weekly injections of a combination of rubidomycin (60 mg/m²) and vincristine (2 mg/m²).

When the glandular enlargement disappeared, the patients were placed on a maintenance treatment combining 6-mercaptopurine and methotrexate, interrupted by periods of reinduction with prednisone, rubidomycin, and vincristine (page 50).

All the cases treated were diffuse forms of the disease with lymphoblasts, with one exception in which the cellular elements seemed rather to be histiocytic.

Our results are shown in Table 55.

We had seven complete remissions, four incomplete remissions, and four failures.

Table 55. *Lymphosarcoma treated by the triple combination of rubidomycin, vincristine, and prednisone (Protocol 06 LA 66). — Results obtained in relation to the stage of the disease*

	No. of patients	Results		
		Complete remissions	Incomplete remissions	Failures
Localized forms I				
Regional forms II	5	3		2
Generalized forms				
— hematological extension	5	2	2	1
— digestive tract extension	2	1		1
— mammary gland extension	1	1		
— renal tract extension	2		2	
Total	15	7	4	4

γ) Complete Remissions

Among the seven complete remissions, three occurred in localized but serious forms of the disease, four in generalized forms.

The Localized Forms

In two cervicomediastinal forms treatment rapidly brought about disappearance of the enlargement of the cervical glands and of bulky mediastinal glands that were responsible for a caval compression syndrome and, in one case, a pleural effusion.

In another patient with bulky cervical glands, a local relapse after radiotherapy and vincaleucoblastine caused a peripheral facial paralysis.

In these three cases the remission was rapid. It occurred towards the 14th day, after 14 days of treatment with corticosteroids and two infusions with rubidomycin and vincristine. In two cases the remission was still continuing six and nine months afterwards. In one case a relapse occurred in the eleventh month and the patient died suddenly at his home shortly afterwards.

The Forms with Visceral Involvement

One remarkable remission was that of a patient with gastric localization and iliolumbar glandular involvement. The remission occurred rapidly, the radiographs showing a considerable improvement as early as the second week of treatment. After a year, the radiographs still showed the patient normal.

In one patient who had an enlarged spleen, involvement of the bone marrow, and multiple foci in the central nervous system, remission for five months was obtained, to be followed by sudden relapse affecting the blood picture.

Finally, with two patients in whom induction treatment had led only to an incomplete remission, remission became complete at reinduction after the first month. In one of the two relapse occurred after five months, in the other the remission was still continuing after nine months. Because of this aged patient's intolerance of prednisone and vincristine, the maintenance treatment was changed after the fifth month.

δ) Incomplete Remissions

The incomplete remissions were characterized by their briefness, relapse occurring even before the reinduction on the 15th day.

ε) Failures

The failures were 1. a patient with lymphosarcoma of the ileocecal valve spreading to the mesentery that was resistant both to drugs and to radiotherapy; 2. two subdiaphragmatic forms of the disease; and 3. a form that when the combination of vincristine, rubidomycin, and prednisone was tried was rapidly progressive, in spite of previous treatment with cobalt and cytoxan, and showed a massive involvement of the bone marrow.

ζ) Tolerance

In general the clinical and hematological tolerance was excellent. In one case alone, already mentioned, we were compelled to modify the maintenance treatment and the reinduction, vincristine having caused polyneuritis in the 75-year old female patient concerned and prednisone at each reinduction causing serious lung conditions. In one case reversible hypoplasia was observed, the invasion of the blood being considerable. In the other patient we mentioned who died at his home twelve months after treatment started, the signs and symptoms described by the family do not exclude the hypothesis of a sudden heart attack.

Other points that deserve emphasis are the rarity and the relative benignity of the marrow aplasia.

η) Discussion

The modes of development of the lymphosarcomas are far from well known, so that comparisons between the various methods of treatment are difficult. However, the favorable effect for several months obtained in patients whose short-term prognosis seemed gloomy speaks well for the combination of drugs that we have described.

The rapidity of relapse in some cases, however, and on the other hand the excellent clinical and hematological tolerance suggest that it might be of value to add to the combination described other drugs of different mode of action, particularly the alkylating agents. A study of this kind is in progress.

b) Reticulosarcoma

There is very little literature on this subject. At an international symposium held in March 1967 in Paris, J. S. BOURDIN reported the treatment of seven cases of reticulosarcoma with rubidomycin [37]. He obtained four favorable results and three partial remissions. TAN et al. also obtained a good result with two children whom they treated and a complete remission in an adult with reticulosarcoma and marrow involvement—a remission that lasted more than eight months.

In view of the usual refractoriness of this kind of malignant hemopathy to drug treatment, it is possible that rubidomycin may be of value in attack treatment or in reinductions. It may be of even greater value in multiple drug treatment.

3. Hodgkin's Disease

As we have treated only six patients with rubidomycin, our experience is very limited. The patients were all in stages III and IV and had become refractory to most of the drugs employed in treatment; they were also in a very bad general state. In one case we obtained complete remission, clinically, radiologically, and biochemically, after treating the patient with a total dose of 12 mg/kg in one month, administered in intermittent injections of 1 mg/kg on three days in each week. In another case a short period of improvement was noted.

Other teams of workers have carried out larger studies. MATHÉ et al. [2, 17] obtained one complete remission lasting two months, four incomplete remissions, two non-total failures, and 13 total failures out of 20 cases treated. No selective effect of rubidomycin was observed on any particular symptom in that study, which was also of a limited series. Two of the patients had localization of the disease in the liver, both having the characteristic biochemical disturbances and one having an enlarged liver and jaundice. Under the influence of the rubidomycin, the liver enlargement, the jaundice, and the biochemical signs of liver infiltration regressed.

4. Myeloma

We have personally treated four patients, but this number is too small to draw any conclusions. In one patient, however, the effect on the pain was rapid and marked. In another case rubidomycin was prescribed after treatment with a combination of drugs had brought about an incomplete remission, as evidenced by disappearance of the pain and an improvement in the protein electrophoretic picture.

Twelve mg/kg of rubidomycin was administered in six weeks and the functional improvement continued, the erythrocyte sedimentation rate, which had still been abnormal after administration of the combination of drugs (60/90), dropped to 30/60, and the fall in the gamma globulins went on. In this case the incomplete remission was still continuing after more than a year on maintenance treatment with chlorambucil.

In recent literature on the subject we have not seen any description of cases of myeloma treated with rubidomycin or daunomycin.

5. Solid Tumors

a) Neuroblastoma

The first therapeutic trials in the United States were concerned with cancer.

The initial results were unsatisfactory in respect of the epitheliomas, but by no means negligible in respect of the neuroblastomas. C. TAN et al. [214] reported interesting results on the neuroblastomas: of 15 cases in children, six were incomplete remissions lasting 1—5 months, two were incomplete remissions lasting less than one month, five were objectively improved in terms of pain, and one was improved in terms of the volume of the tumor. Only one case was a total failure.

O. SCHWEISGUTH [150] obtained one complete remission lasting a month and one non-total failure in six cases.

b) Rhabdomyosarcoma

Out of 10 cases of embryonic rhabdomyosarcoma treated in children, C. TAN [214] observed three regressions of the tumor.

c) Cancer

The other solid tumors seem to be resistant to rubidomycin. C. TAN [214] obtained no therapeutic response in 13 adult cancers. In 10 cases of cancer of the head and neck we noted only one complete regression of the tumor, an ulcerating cancer of the buccal floor (T4—N2) in a woman of 81 years of age. The ulcer healed, the enlargement of the glands disappeared, and there remained only a slight induration at the initial site of the tumor, underneath a mucosa that had regained a strictly normal microscopic appearance. In this case 7 mg/kg of rubidomycin was administered in 15 days.

Because the number of patients treated is still small, no formal conclusions can be drawn about the effect of rubidomycin. It is not impossible that beneficial effects may be observed on some of the epitheliomas that were by chance not represented in the first series studied. Nor is it impossible that rubidomycin may be of value in some combinations of drugs used to treat epitheliomas. At present, however, it would be better to expect that rubidomycin will be ineffective in such treatment. There is a striking contrast, with rubidomycin as with many other antimitotic drugs, between its range of action on the malignant blood diseases, especially certain leukemias, and its total lack of action on the epitheliomas.

d) Intra-arterial Route

Rubidomycin has been used by the intra-arterial route in doses of 1 mg/kg daily for 5—10 days. The studies in progress are not sufficiently advanced for any results to be reported here.

Conclusions

I. Rubidomycin, an antibiotic with antitumoral activity, is extracted from several strains of *Streptomyces coeruleorubides*.

Daunomycin, extracted from *Streptomyces peucetius*, is identical with rubidomycin.

II. Rubidomycin belongs to the anthracycline group (glycoside antibiotics) and occurs as a red crystalline powder.

Its full formula has been worked out almost completely by acid hydrolysis; it is the glycoside of daunomycin and daunosamine.

III. Experimental study shows the antitumoral activity of rubidomycin against solid tumors (sarcomas and carcinomas), Ehrlich's ascitic tumor, leukemia, and tumors of viral origin. It is one of the first antibiotics to be active against certain grafted mice leukemias: AKR and C 1498 leukemias, acute lymphoblastic leukemia X grafted on strain C 57 Bl/6, and L 1210 leukemia.

IV. Experimentally in mice and rats, acute and subacute toxicity $(LD/_{50})$ lies between 12.5 and 25 mg/kg body weight. Long-term toxicity has been studied in the rabbit and dog: the maximum tolerated daily doses of rubidomycin administered intravenously during a course of treatment for three months approximate 0.25 mg/kg body weight. This dose may cause testicular changes, particularly in the dog, and myocardial attacks in the rabbit.

V. Analysis of the mechanism of action of rubidomycin has shown strong *in vitro* mitostatic activity in low doses. This activity changes the chromosome structure and disturbs the metabolism first of DNA and then of RNA. The cause of the activity does not seem to be the action of rubidomycin on cell energy metabolism. Rubidomycin is an immunodepressive agent.

VI. Rubidomycin has been used in the treatment of leukemia, sarcoma, and certain epitheliomas in man.

The average dose is 0.5—2 mg/kg daily (15—70 mg/m²), administered intravenously. The injections are given in different schedules according to the indications—daily in the treatment of acute leukemia, every 2—3 days in the treatment of the other malignant blood diseases and of cancer. The total cumulative dose should not exceed 25—30 mg/kg (750—900 mg/m²), administered over a period of six months.

VII. Rubidomycin causes in man three kinds of complication: 1. marrow aplasias, which are necessary precursors of remission in treatment, so that the risk of their occurring must be taken deliberately and means provided for energetic restoration of the blood picture before treatment is started; 2. cardiac mishaps, which are especially to be feared when the total cumulative dose exceeds 25—30 mg/kg (750—900 mg/m²); and finally, 3. various other minor complications. So far we have not observed any liver or kidney toxicity.

VIII. The principal properties of rubidomycin used alone in the treatment of acute leukemia are as follows: very great speed of action, absence of cross-resistance between the drug and the other active preparations (cortisone, vincristine, folic acid antagonists, 6-mercaptopurine, cytosine arabinoside, methyl-GAG), and particularly activity against all the cell types of acute leukemia—lymphoblasts, myeloblasts, and promyelocytes. On the other hand, the use of the drug is made difficult because of the need to obtain a rapid phase of aplasia before the remission and the great difference between patients in susceptibility to the drug. If the treatment goes on too long it causes a fatal aplasia, and if it is too short it will not only be ineffective but will also allow cells resistant to rubidomycin to develop.

IX. Using rubidomycin alsone, 60% of complete remissions can be obtained in acute lymphoblastic leukemia, both in cases not previously treated and in cases refractory to other forms of treatment.

In this type of acute leukemia, rubidomycin brings about the most remarkable morphological changes in the lymphoblasts. Frequent and close cytological examination of the marrow picture is therefore of help in an assessment of the drug's effect.

X. Rubidomycin is the first drug to enable a high percentage of complete remissions to be obtained in acute myeloblastic leukemia. In a homogeneous series of 64 cases it was slightly higher than 50%. Other studies in progress at present do not seem to give as high a percentage of remissions; in them the frequency is between 35% and 55% for remissions obtained, on condition that treatment is continuous from the start. This method involves the risk of severe aplasia and requires the application of very energetic measures to restore the blood picture.

XI. Acute promyelocytic leukemia is also very sensitive to rubidomycin. With it the very special symptoms of this disease (profuse bleeding and fibrinolysis) are quickly attenuated and remissions are obtained.

XII. It is now standard practice to use combinations of drugs in the induction of complete remissions in acute leukemia.

With reinductions the period of remission can be very markedly increased in patients whose maintenance treatment includes one or more antimetabolites. On the basis of this concept, we have employed rubidomycin in these drug combinations and for reinductions.

XIII. In the treatment of acute lymphoblastic leukemia we used the combination of prednisone, vincristine, and rubidomycin in inductions and reinductions and 6-mercaptopurine and methotrexate in maintenance treatment.

We obtained a very high percentage (90%) of complete remissions: 85 complete remissions out of 87 children treated (97%), and 32 complete remissions out of 43 adults (74%). This combination with rubidomycin brought about a quick remission in most cases (between the first and the third week), and remission was also obtained in very unfavorable cases such as the tumoral and hyperleukocytic forms of the disease. Not enough time has elapsed to make it possible to assess the distribution of the relapses over time.

XIV. The following combination of drugs is under trial in the treatment of acute myeloblastic leukemia: cytosine arabinoside, methyl-glyoxal-bis-guanyl-hydrazone, 6-mercaptopurine, rubidomycin, and prednisone. The combination seems to be satisfactorily tolerated.

XV. Rubidomycin is indicated in the treatment of chronic myeloid leukemia in the following conditions: in cases in which a rapid effect is desirable, as for example for patients with thrombosis of the corpus cavernosum and priapism or with an infarct of the spleen; in forms of the disease at the myelocytic phase that are refractory to Busulfan; and finally, in attack treatment or in reinductions in combination with other treatment.

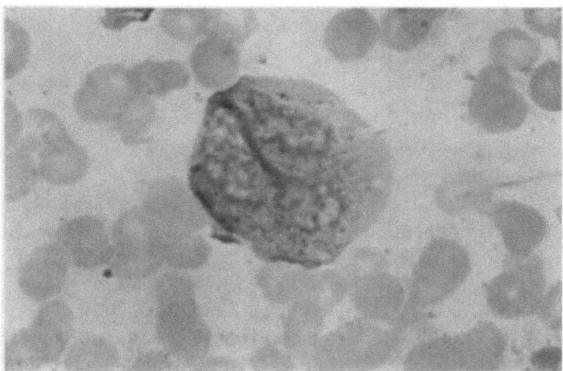

Fig. 66. Myeloblasts in acute myeloblastic leukemia, little changed by treatment

XVI. Rubidomycin has no effect on chronic lymphoid leukemia.

XVII. In the treatment of lymphosarcoma, the same combination of prednisone, rubidomycin, and vincristine as is used in the treatment of acute leukemia gave reasonable results in our limited series of control group patients.

XVIII. On the other hand, rubidomycin employed alone is ineffective or poorly effective in the treatment of Hodgkin's disease, reticulosarcoma, and cancer. It may, however, have a place in multiple drug therapy if an antimitotic antibiotic is to be used. Some improvement has been obtained in neuroblastoma, but it has been very inconstant.

References

1. Acute Leukemia Group B: New treatment schedule with improved survival in childhood leukemia. J. Amer. med. Assoc. **164**, 75—81 (1965).
2. AMIEL, J. L., L. SCHWARZENBERG, M. SCHNEIDER, A. CATTAN, J. R. SCHLUMBERGER, M. HAYAT et G. MATHE: Valeur de la rubidomycine dans le traitement de la maladie de Hodgkin. Colloque international sur la rubidomycine et la daunomycine. Paris, Hôpital Saint-Louis, 11 mars 1967. Pathol. Biol. **15**, 961—962 (1967).
3. ARCAMONE, F.: La struttura della daunomicina. Chim. Ind. (Milan) **47**, 773 (1965).
4. — La constitution chimique de la daunomycine. Colloque international sur la rubidomycine et la daunomycine. Paris, Hôpital Saint-Louis, 11 mars 1967. Pathol. Biol. **15**, 893—895 (1967).
5. —, G. CASSINELLI, P. OREZZI, G. FRANCESCHI, and R. MONDELLI: Daunomycin. II. The structure and stereochemistry of daunosamine. J. Amer. chem. Soc. **86**, 5335—5336 (1964).
6. —, A. DI MARCO, M. GAETANI e T. SCOTTI: Isolamento ed attività antitumorale di un antibiotico da Streptomyces sp. Giorn. Microbiol. **9**, 83—90 (1961).
7. —, G. FRANCESCHI, P. OREZZI, G. CASSINELLI, W. BARBIERI, and R. MONDELLI: Daunomycin. I. The structure of daunomycinone. J. Amer. chem. Soc. **86**, 5334—5335 (1964).
8. ASHESHOV, I. N., and J. J. GORDON: Rutilantin: an antibiotic substance with antiphage activity. Biochem. J. **81**, 101—104 (1961).
9. BACH, F., and K. HIRSCHHORN: Lymphocyte interaction: a potential histocompatibility test in vitro. Science **143**, 813—814 (1964).
10. BAIN, B., M. R. VAS, and L. LOWENSTEIN: The development of large immature mononuclear cells in mixed leucocyte cultures. Blood **23**, 108—115 (1964).
11. BARBIERI, P., A. DI MARCO, R. MAZZOLENI, M. MENOZZI e A. SANFILIPPO: Azione della daunomicina sul metabolismo degli acidi nucleinici nell' E. coli B. Giorn. Microbiol. **12**, 71—82 (1964).
12. BELOVA, I. P.: Histopathological changes in organs and tissues of experimental animals after rubomycin administration (russ.). Antibiotiki **11**, 133—139 (1966).
13. BERENBAUM, M. C.: Effect of cytotoxic agents on antibody production. Nature **185**, 167—168 (1960).
14. BERNARD, J.: Maladies du sang et des organes hématopoïétiques. Paris: Flammarion éd., 1948.
15. — Conférence sur la chimiothérapie de la maladie de Burkitt (U.I.C.C.). Kampala. Janvier 1966.
16. — Traitements actuels des leucémies aiguës lymphoblastiques. Effets de la méthode de réinduction. Nouveaux médicaments (rubidomycine). Entretiens de Bichat-Thérapeutique. 1967. Paris. Expansion Scientifique Française éd., 97—99 (1967). Semaine Ther., Sem. Hôp. **44**, 308—310 (1968).
17. —, et M. BOIRON: Les leucémies à promyélocytes. Nouv. Rev. franç. Hémat. **4**, 11—14 (1964).
18. — —, CL. JACQUILLAT, Y. NAJEAN, M. SELIGMANN, J. TANZER et M. WEIL: Traitements actuels des leucémies aiguës. Presse méd. **74**, 1241—1245 (1966).
19. — — — et M. WEIL: Premiers résultats de l'association du méthylglyoxal bis(guanylhydrazone) et de la 6-mercaptopurine dans le traitement des leucémies aiguës de la série granulocytaire. Presse méd. **72**, 807—809 (1964).
20. — — — — et Y. NAJEAN: Un nouvel agent actif dans le traitement des leucémies aiguës, la cytosine-arabinoside. Presse méd. **74**, 799—802 (1966).

21. BERNARD, J., M. BOIRON, CL. JACQUILAT, Y. NAJEAN, M. WEIL et M. THOMAS: Traitement des leucémies aiguës lymphoblastiques de première atteinte par une association de prednisone-vincristine-rubidomycine. Colloque international sur la rubidomycine et la daunomycine. Paris, Hôpital Saint-Louis, 11 mars 1967. Pathol. Biol. 15, 919—920 (1967).

22. — —, P. LORTHOLARY, and J. P. LEVY: The very acute leukemias. Cancer Res. 25. 1675 to 1676 (1965).

23. — —, A. MANUS, J. P. LEVY, and J. LELLOUCH: Factors influencing survival time in patients with acute leukemia. Nat. Cancer Inst. Monograph. 15, 359—365 (1964).

24. — —, M. WEIL, J. P. LEVY, M. SELIGMANN et Y. NAJEAN: Etude de la rémission complète des leucémies aiguës (analyse de 300 observations). Nouv. Rev. franç. Hémat. 2, 195—222 (1962).

25. —, et CL. JACQUILLAT: La rubidomycine. Nouv. Rev. franç. Hémat. 7, 317—319 (1967).

26. — —, M. BOIRON, Y. NAJEAN, M. SELIGMANN, J. TANZER, M. WEIL et P. LORTHOLARY: Essai de traitement des leucémies aiguës lymphoblastiques et myéloblastiques par un antibiotique nouveau: la rubidomycine (13057 R.P.). Etude de 61 observations. Presse méd. 75, 951—955 (1967).

27. — — — — — et M. WEIL: Les très longues rémissions complètes des leucémies aiguës. Presse méd. 73, 457—459 (1965).

28. —, et G. MATHE: Etude des leucoses aiguës de l'enfance traitées par l'association antifoliques-cortisone. Sang. 23, 12—27 (1952).

29. — —, J. BOULAY, B. CEOARA et J. CHOME: La leucose aiguë à promyélocytes. Etude portant sur 20 observations. Schweiz. med. Wschr. 89, 604—608 (1959).

30. —, et M. SELIGMANN: Le traitement des leucoses par la 6-mercaptopurine. Sem. Hôp. Paris 30, 2971—2977 (1954).

31. BERTALANFFY, L. VON, and I. BICKIS: Identification of cytoplasmic basophilia (ribonucleic acid) by fluorescence microscopy. J. Histochem. Cytochem. 4, 481—493 (1956).

32. —, H. MASIN, and F. MASIN: A new and rapid method for diagnosis of vaginal and cervical cancer by fluorescence microscopy. Cancer 11, 873—887 (1958).

33. BESSIS, M., et J. BERNARD: Remarquables résultats du traitement par exsanguinotransfusion d'un cas de leucémie aiguë. Bull. Mém. Soc. Méd. Hôp., Paris, 63, 28—29 et 871—876 (1947).

34. BLUMBERG, N. A., and N. I. KALAMOVA: Antiviral activity of rubomycin (russ.). Antibiotiki 12, 898—903 (1967).

35. BOIRON, M., CL. JACQUILLAT, M. WEIL, and J. BERNARD: Combination of methylglyoxal bis(guanylhydrazone) [NSC-32946] and 6-mercaptopurine [NSC-755] in acute granulocytic leukemia. Cancer Chemotherapy Rept. 45, 69—73 (1965).

36. — — —, M. THOMAS et J. BERNARD: Traitement des leucémies aiguës granulocytaires par la rubidomycine. Colloque international sur la rubidomycine et la daunomycine. Paris, Hôpital Saint-Louis, 11 mars 1967. Pathol. Biol. 15, 921—924 (1967).

37. BOURDIN, J. S., B. CLAVEL, R. T. SARACINO, H. HEBERT et I. LILLE: Traitement des hémato-sarcomes par la rubidomycine. Colloque international sur la rubidomycine et la daunomycine. Paris, Hôpital Saint-Louis, 11 mars 1967. Pathol. Biol. 15, 963—965 (1967).

38. BRAZHNIKOVA, M. G., N. V. KONSTANTINOVA, V. A. POMAZKOVA, and B. V. ZAKHAROV: Physiochemical properties of the antitumor antibiotic rubomycin, produced by Act. coeruleorubidus (russ.). Antibiotiki 11, 763—767 (1966).

39. BRAZHNIKOVA, M. G., I. N. KOVSHAROVA, N. V. KONSTANTINOVA, and V. A. POMAZKOVA: Physicochemical properties of the antitumor antibiotic rubomycin, produced by Actinomyces coeruleorubidus. International IX Congress for Microbiology, Moscow 24—30, 7, 1966. Abstracts of Papers, p. 167. Organizing Committee of the Congress, Moscow (1966).

40. BROCKMANN. H.: Anthracyclinone und Anthracycline (Rhodomycinone, Pyrromycinone und ihre Glucoside). Fortschr. Chem. Org. Naturstoffe 21, 121—182 (1963).

41. —, u. K. BAUER: Rhodomycin, ein rotes Antibioticum aus Actinomyceten. Naturwissenschaften 37, 492—493 (1950).

42. —, u. N. GRUBHOFER: Zur Kenntnis des Actinomycins C. Naturwissenschaften 37, 494 bis 496 (1950).

43. BROCKMANN, H., u. W. LENK: Über Actinomycetenfarbstoffe. VII. Pyrromycin. Chem. Ber. **92**, 1904—1909 (1959).
44. —, u. P. PATT: Iso-rhodomycin A, ein neues Antibioticum aus Streptomyces purpurascens, Rhodomycine. III. Mitteil.; Antibiotica aus Actinomyceten, XXXII. Mitteil. Chem. Ber. **88**, 1455—1468 (1955).
45. —, E. SPOHLER u. T. WAEHNELDT: Rhodosamin, Isolierung, Konstitution und Konfiguration. Chem. Ber. **96**, 2925—2936 (1963).
46. —, u. E. WIMMER: Die Konstitution des β-, γ-, ε- und ζ-Rhodomycinons. Rhodomycine (VIII); Antibiotica aus Actinomyceten (LI). Chem. Ber. **98**, 2797—2804 (1965).
47. BRUBAKER, C. A., H. E. WHELER, M. J. SONLEY, C. B. HYMAN, K. O. WIILIAMS, and D. HAMMOND: Cyclic chemotherapy for acute leukemia in children. Blood **22**, 820 to 821 (1963).
48. BURCHENAL, J. H.: Long-term remissions of acute leukemia spontaneous and induced. Scand. J. Haematol. Series II haematologica I, 4756 (1965).
49. —, and M. L. MURPHY: Long-term survivors in acute leukemia. Cancer Res. **25**, 1491 to 1494 (1965).
50. CAHN, R. S., C. K. INGOLD, and V. PRELOG: The specification of asymmetric configuration in organic chemistry. Experientia **12**, 81—94 (1956).
51. CALENDI, E., A. DI MARCO, M. REGGIANI, B. SCARPINATO, and L. VALENTINI: On physicochemical interactions between daunomycin and nucleic acids. Biochim. Biophys. Acta **103**, 25—49 (1965)
52. CASSINELLI, G., et P. OREZZI: La daunomicina: un nuovo antibiotico ad attività citostatica. Isolamento e proprietà. Giorn. Microbiol. **11**, 167—174 (1963).
53. CHABBERT, Y., and H. VIAL: Cytotoxic substances in monolayer tissue cultures by an agar diffusion method. Exp. Cell Res. **22**, 264—274 (1961).
54. CHANCE, B., and B. HESSE: Metabolic control mechanisms. I. Electron transfer in the mammalian cell. J. Biol. Chem. **234**, 2404—2412 (1959).
55. CHAUVERGNE, J., et CL. LAGARDE: Chimiothérapie anticancéreuse par la rubidomycine. (Note préliminaire sur 14 observations). Colloque international sur la rubidomycine et la daunomycine. Paris, Hôpital Saint-Louis, 11 mars 1967. Pathol. Biol. **15**, 969—970 (1967).
56. CHEVALIER, L., and O. GLIDEWELL (Acute leukemia group B): Schedule of 6-mercaptopurine and effect of inducer drugs in prolongation of remission maintenance in acute leukemia. Proc. Amer. Ass. Cancer Res. **8**, 10, abstr. 37 (1967).
57. Children's Cancer Group A: Preliminary evaluation of daunomycin in children with acute leukemia. Colloque international sur la rubidomycine et la daunomycine. Paris, Hôpital Saint-Louis, 11 mars 1967. Pathol. Biol. **15**, 939 (1967).
58. Colloque International sur la Rubidomycine et la Daunomycine. Paris, Hôpital Saint-Louis, 11 mars 1967. Pathol. Biol. **15**, 887—972 (1967).
59. COSTA, G., e G. ASTALDI: Effecto della daunomicina sull' attività proliferativa di cellule staminali ottenute in coltura da sangue umano normale. Tumori **50**, 477—480 (1964).
60. — — Inhibition by daunomycin of the stem cell development of lymphocytes. Gazz. Intern. Med. Chir. **70**, 597—600 (1965).
61. DAMESHEK, W., and J. MITUS: Seven-years' remission in an adult with acute leukemia. New Engl. J. Med. **268**, 870—873 (1963).
62. DESPOIS, R., M. DUBOST, D. MANCY, R. MARAL, L. NINET, S. PINNERT, J. PREUD'HOMME, Y. CHARPENTIE, A. BELLOC, N. DE CHEZELLES, J. LUNEL et J. RENAUT: Un nouvel antibiotique doué d'activité antitumorale: la rubidomycine (13057 R.P.). I. Préparation et propriétés. Arzneimittel-Forsch. **17**, 934—939 (1967).
63. —, M. DUBOST, D. MANCY, R. MARAL, L. NINET, S. PINNERT, J. PREUD'HOMME, Y. CHARPENTIE, A. BELLOC, N. DE CHEZELLES, J. LUNEL et J. RENAUT: Isolement d'un nouvel antibiotique doué d'activité antitumorale: la rubidomycine (13057 R.P.). Identité de la rubidomycine et de la daunomycine. Colloque international sur la rubidomycine et la daunomycine. Paris, Hôpital Saint-Louis, 11 mars 1967. Pathol. Biol. **15**, 887—891 (1967).
64. DI MARCO, A.: Daunomycin pharmacological activity at the cellular level. Colloque international sur la rubidomycine et la daunomycine. Paris, Hôpital Saint-Louis, 11 mars 1967. Pathol. Biol. **15**, 897—902 (1967).

65. DI MARCO, A.: Attività biologica ed utilizzazione terapeutica dell' antibiotico daunomicina. Tumori **53**, 269—291 (1967).

66. — Mechanism of action of daunomycin and related antibiotics. In: Mechanism of action of antibiotics. Ed. D. GOTTLIEB, and P. D. SHAW. Springer-Verlag Heidelberg-New York: 1967.

67. —, G. BORETTI e A. RUSCONI: Trasformazione metabolico della daunomicina da parte di estratti di tissuti. Farmaco (Pavia), Ed. Sci. **22**, 535—542 (1967).

68. —, M. GAETANI, L. DORIGOTTI, M. SOLDATI e O. BELLINI: Studi sperimentali sull' attività antineoplastica del nuovo antibiotico daunomicina. Tumori **49**, 203—217 (1963).

69. — — — — — Daunomycin: a new antibiotic with antitumor activity. Cancer Chemotherapy Rept. **38**, 31—38 (1964).

70. — —, P. OREZZI, B. SCARPINATO, R. SILVESTRINI, M. SOLDATI, T. DASDIA, and L. VALENTINI: "Daunomycin", a new antibiotic of the rhodomycin group. Nature **201**, 706—707 (1964).

71. — — —, and M. SOLDATI: Antitumor activity of a new antibiotic: daunomycin. IIIrd International Congress of Chemotherapy, Stuttgart 1963. Stuttgart: Georg Thieme 1023—1031 (1964).

72. —, R. SILVESTRINI, S. DI MARCO, and T. DASDIA: Inhibiting effect of the new cytotoxic antibiotic daunomycin on nucleic acids and mitotic activity of HeLa cells. J. Cell Biol. **27**, 545—550 (1965).

73. —, M. SOLDATI, A. FIORETTI e T. DASDIA: Ricerche sull' attività della daunomicina su cellule normali e neoplastiche coltivate in vitro. Tumori **49**, 235—252 (1963).

74. — — — — Activity of daunomycin, a new antitumor antibiotic on normal and neoplastic cells grown in vitro. Cancer Chemotherapy Rept. **38**, 39—47 (1964).

75. DI PIETRO, S., e L. GENNARI: Chemioterapia antiblastica regionale dei tumori del distretto cervico-facciale mediante infusione endoarteriosa continua. Tumori **50**, 267—308 (1964).

76. DORIGOTTI, L.: Studio al microscopio elletronico delle modificazioni indotte dalla daunomicina sulle cellule HeLa. Tumori **50**, 117—135 (1964).

77. DOROZHINSKY, V. B.: The use of polarographic method determination of rubomycin concentrations in solutions (russ.). Antibiotiki **12**, 795—797 (1967).

78. DREYFUS, B., C. SULTAN et P. BESSON: Note sur le traitement de six leucémies aiguës myéloblastiques par la rubidomycine. Colloque international sur la rubidomycine et la daunomycine. Paris, Hôpital Saint-Louis, 11 mars 1967. Pathol. Biol. **15**, 925—927 (1967).

78a — —, M. BOIRON, CL. JACQUILLAT, M. WEIL et H. ROCHANT: Sur le traitement d'attaque par la rubidomycine de 19 cas de leucémies aiguës myéloblastiques. Presse méd. **76**, 55—57 (1968).

79. DUBOST, M., P. GANTER, R. MARAL, L. NINET, S. PINNERT, J. PREUD'HOMME et G. H. WERNER: Un nouvel antibiotique à propriétés cytostatiques: la rubidomycine. C. R. Acad. Sci. Paris **257**, 1813—1815 (1963).

80. — — — — — — — Rubidomycin: a new antibiotic with cytostatic properties. Cancer Chemotherapy Rept. **41**, 35—36 (1964).

81. DUDNIK, Y. V., and G. F. GAUSE: On the mechanism of rubomycin action (russ.). Antibiotiki **12**, 17—22 (1967).

82. *Editorial:* Rubidomycin in acute leukaemia. Brit. med. J. **1967 II**, 587—588.

83. ELLISON, R. R., J. F. HOLLAND, R. T. SILVER, J. BERNARD, and M. BOIRON: Cytosine arabinoside, a new drug for induction of remissions in acute leukemia. IXth International Cancer Congress, Tokyo, oct. 1966. Abstracts of Paper. Section II, 15-C, leukemia II, 645.

84. ETTLINGER, L., E. GÄUMANN, R. HÜTTER, W. KELLER-SCHIERLEIN, F. KRADOLFER, L. NEIPP, V. PRELOG, P. REUSSER u. H. ZÄHNER: Stoffwechselprodukte von Actinomyceten. XVI. Cinerubine. Chem. Ber. **92**, 1867—1879 (1959).

85. FARBER, S., G. D'ANGIO, A. EVANS, and A. MITUS: Clinical studies of actinomycin D with special reference to Wilms' tumor in children. Ann. N. Y. Acad. Sci. **89**, 421—425 (1960).

86. FARBER, S., L. K. DIAMOND, R. D. MERCER, R. F. SYLVESTER, JR., and J. A. WOLFF: Temporary remissions in acute leukemia in children produced by folic acid antagonist, 4-aminopteroylglutamic acid (aminopterin). New Engl. J. Med. **238**, 787—793 (1948).

87. FARMITALIA: Italian Patent Specification N⁰ 29060 filed 16/11/62.

88. FREI, E. III, E. J. FREIREICH, E. GEHAN, D. PINKEL, J. F. HOLLAND, O. S. SELAWRY, F. HAURANI, C. L. SPURR, D. M. HAYES, G. W. JAMES, H. ROTHBERG, D. B. SODEE, R. W. RUNDLES, L. R. SCHROEDER, B. HOOGSTRATEN, I. J. WOLMAN, D. G. TRAGGIS, T. COOPER, B. R. GENDEL, F. G. EBAUGH, and R. J. TAYLOR: Studies of sequential and combination antimetabolite therapy in acute leukemia: 6-mercaptopurine and methotrexate. Blood **18**, 431—454 (1961).

89. —, M. KARON, R. H. LEVIN, E. J. FREIREICH, R. J. TAYLOR, J. HANANIAN, O. S. SELAWRY, J. F. HOLLAND, B. HOOGSTRATEN, I. J. WOLMAN, E. ABIR, A. SAWITSKY, S. LEE, S. D. MILLS, E. O. BURGERT, C. L. SPURR, R. B. PATTERSON, F. G. EBAUGH, G. W. JAMES III, and J. H. MOON: The effectiveness of combinations of antileukemic agents in inducing and maintaining remission in children with acute leukemia. Blood **26**, 642—656 (1965).

90. —, R. H. LEVIN, G. P. BODEY, E. E. MORSE, and E. J. FREIREICH: The nature and control of infections in patients with acute leukemia. Cancer Res. **25**, 1511—1515 (1965).

91. FREIREICH, E. J.: Acute leukemia. Med. Ann. District of Columbia **31**, 675—680 (1962).

92. — Effectiveness of platelet transfusion in leukemia and aplastic anemia. Transfusion (Phil.) **6**, 50—54 (1966).

93. —, E. GEHAN, E. FREI III, L. R. SCHROEDER, I. J. WOLMAN, R. ANBARI, E. D. BURGERT, S. D. MILLS, D. PINKEL, O. S. SELAWRY, J. H. MOON, B. R. GENDEL, C. L. SPURR, R. STORRS, F. HAURANI, B. HOOGSTRATEN, and S. LEE: The effect of 6-mercaptopurine on the duration of steroid induced remissions in acute leukemia. A model for evaluation of other potentially useful therapy. From the acute leukemia group B. Blood **21**, 699—716 (1963).

94. —, G. JUDSON, and R. H. LEVIN: Separation and collection of leukocytes. Cancer Res. **25**, 1516—1520 (1965).

95. —, M. KARON, F. FLATOW, and E. FREI III: Effect of intensive cyclic chemotherapy (BIKE) on remission duration in acute lymphocytic leukemia. Proc. Amer. Ass. Cancer Res. **6**, 20, abstr. 76 (1965).

96. — —, and E. FREI III: Quadruple combination therapy (VAMP) for acute lymphocytic leukemia of childhood. Proc. Amer. Ass. Cancer Res. **5**, 20, abstr. 76 (1964).

97. —, R. H. LEVIN, J. WHANG, P. CARBONE, W. BROSON, and E. E. MORSE: The function and rates of transfused leukocytes from donors with myelocytic leukaemia in leucopenic recipients. Ann. N. Y. Acad. Sci. **113**, 1081 (1964).

98. GAUSE, G. F.: Mechanism of the biochemical action of antibiotics (russ.). Antibiotiki **9**, 946—950 (1964).

99. — Biochemical mechanisms involved in the action of antineoplastic antibiotics with reference to the problem of research on new active agents (russ.). Vest. Akad. Med. Nauk SSSR **20**, 46—52 (1965).

100. — Aspects of antibiotic research. Chem. Ind. (London) **36**, 1506—1513 (1966).

101. —, T. P. PREOBRASHENSKAYA, E. S. KUDRINA, N. O. BLINOV, I. D. RJABOWA, and M. A. SWESCHINIKOVA: Problems of classification of actinomycete antagonist. Inst. for Research of New Antibiotics, Acad. Med. Sci., National Press of Medical Literature, Medzig, Moskow, U.S.S.R. (1957). Zur Klassifizierung der Actinomyceten, Fischer G., Jena (1958). English Amer. Inst. Biol. Sci., Washington (1959).

102. —, T. P. PREOBRASHENSKAYA, N. A. MANAPHOVA, V. K. KOVALENKOVA, Y. V. DUDNIK, and G. V. KOCHETKOVA: Antitumor antibiotic rubomycin: production and the mechanism of action. International IX Congress for Microbiology, Moscow, 24—30/7/66. Abstracts of Papers, p. 166. Organizing Committee of the Congress, Moscow (1966).

103. GEORGE, M., and J. H. VAUGHAN: In vitro cell migration as a model for delayed hypersensitivity. Proc. Soc. exp. Biol. Med. **111**, 514—521 (1963).

104. GOLDBERG, I. H.: Mode of action of antibiotics. II. Drugs affecting nucleic acid and protein synthesis. Amer. J. Med. **39**, 722—752 (1965).

105. GOLDBERG, L. E.: Pharmacological study of rubomycin (russ.). Antibiotiki **11**, 126—133 (1966).
106. GOLDIN, A., J. M. VENDITTI, S. R. HUMPHREYS, and N. MANTEL: Modification of treatment schedules in the management of advanced mouse leukemia with amethopterine. J. Nat. Cancer Inst. **17**, 203—212 (1956).
107. — —, I. KLINE, M. GANG, and V. S. WARAVDEKAR: Factors influencing the therapeutic effectiveness of antitumor agents. Vth International Congress of Chemotherapy, Wien (Austria) **3**, 441—447 (1967). Verlag der Wiener Medizinischen Akademie (1967).
108. GREIN, A., e C. SPALLA: Studio sui coremi formati in culture di Streptomyces peucetius. Giorn. Microbiol. **10**, 175—184 (1962).
109. — —, A. DI MARCO e G. CANEVAZZI: Descrizione e classificazione di un attinomicete (Streptomyces peucetius sp. nova) produttore di una sostanza ad attività antitumorale: la daunomicina. Giorn. Microbiol. **11**, 109—118 (1963).
110. HACKMANN, C.: Experimentelle Untersuchungen über die Wirkung von Actinomycin C (HBF 386) bei bösartigen Geschwulsten. Z. Krebsforsch. **58**, 607—613 (1952).
111. — HBF 386 (Actinomycin C), ein cytostatisch wirksamer Naturstoff. Strahlentherapie **90**, 296—300 (1953).
112. HADDOW, A., and G. M. TIMMIS: Myleran in chronic myeloid leukaemia. Chemical constitution and biological action. Lancet I (6753), 207—208 (1953).
113. HANANIAN, J., J. F. HOLLAND, and P. SHEEHE: Intensive chemotherapy of acute lymphocytic leukemia in children. Proc. Amer. Ass. Cancer Res. **6**, 26, abstr. 100 (1965).
114. HARDISTY, R. M., and P. M. NORMAN: Preliminary experience with rubidomycin in the treatment of acute lymphoblastic leukaemia resistant to other antimetabolites. Colloque international sur la rubidomycine et la daunomycine. Paris, Hôpital Saint-Louis, 11 mars 1967. Pathol. Biol. **15**, 941—942 (1967).
115. HARTMANN, G., H. GOLLER, K. KOSCHEL, W. KERSTEN u. H. KERSTEN: Hemmung der DNA-abhängigen RNA- und DNA-Synthese durch Antibiotica. Biochem. Z. **341**, 126—128 (1964—1965).
116. HAYHOE, F. G., D. GUAGLINO, and R. DOLL: The cytology and cytochemistry of acute leukemias, a study of 140 cases. London, Her Majesty's Stationery Office, 102 (1964).
117. HEINEMANN, B., and A. J. HOWARD: Antiphage properties of compounds possessing both antitumor and inducing activities. Antimicrobial Agents and Chemotherapy, 126—130 (1964).
118. — — Effect of compounds with both antitumor and bacteriophage-inducing activities on Escherichia coli nucleic acid synthesis. Antimicrobial Agents and Chemotherapy, 488—492 (1965).
119. HENDERSON, E. S.: Combination chemotherapy of acute leukemia. Vth International Congress of Chemotherapy, Wien (Austria) **3**, 293—297 (1967), Verlag der Wiener Medizinischen Akademie (1967).
120. —, E. J. FREIREICH, M. KARON, and W. ROOSSE: High dose combination chemotherapy in acute lymphocytic leukemia of childhood. Proc. Amer. Ass. Cancer Res. **7**, 30, abstr. 115 (1966).
121. HEWLETT, J. S., J. D. BATTLE, JR., R. C. BISHOP, W. M. FOWLER, S. O. SCHWARTZ, P. S. HAGEN, and J. LOUIS: Phase II study of A-8103 (NSC-25154) in acute leukemia in adults. Cancer Chemotherapy Rept. **42**, 25—28 (1964).
122. HILZ, H., B. HUBMANN, M. OLDEKOP, M. SCHOLZ u. M. GOSSLER: Die Wirkung von Röntgenstrahlen und cytostatischen Verbindungen auf DPN-Gehalt und DNS-Synthese in Ascitestumorzellen. Biochem. Z. **336**, 62—76 (1962).
123. HOLLAND, J. F.: Daunomycin treatment in acute leukemia. Colloque international sur la rubidomycine et la daunomycine. Paris, Hôpital Saint-Louis, 11 mars 1967. Pathol. Biol. **15**, 929—932 (1967).
124. —, E. FREI III, and J. H. BURCHENAL: Criteria for remission in acute leukemia. Proceedings of 6th International Congress of the Int. Soc. Hemat., Boston (1956). New York: Grune and Stratton 1958.
125. HONIG, G. R., M. E. SMULSON, and M. RABINOVITZ: A requirement of RNA synthesis for oxidation-dependent biosynthesis in sarcoma 37 ascites-tumor cells. Biochim. Biophys. Acta **129**, 576—584 (1966).

126. HOWARD, J. P., and C. TAN: Combined daunomycin-prednisone inductions in acute leukemia. Proc. Amer. Ass. Cancer Res. **8**, 32, abstr. 124 (1967).

127. HUGULEY, C. M., W. R. VOGLER, J. W. LEA, C. C. CORLEY, and M. E. LOWREY: Acute leukemia treated with divided doses of methotrexate. Arch. Internal Med. **115**, 23—28 (1965).

128. JACQUILLAT, CL.: Antimitotic used for leukemia. Med. Tribune **8**, 20 (1967).

129. —, M. BOIRON, Y. NAJEAN, M. WEIL et J. BERNARD: Traitement d'attaque et d'entretien des leucémies aiguës lymphoblastiques, effets de la méthode dite de «réinductions» systématiques. Marseille Méd. **104**, 1—11 (1967).

130. — —, M. WEIL, A. MAZELIER et J. BERNARD: Résultats du traitement des hémopathies malignes par la rufocromomycine (5278 R.P.). A propos de 57 observations. Presse Méd. **73**, 2003—2006 (1965).

131. — —, M. WEIL, J. TANZER, Y. NAJEAN, and J. BERNARD: Rubidomycin. A new agent active in the treatment of acute lymphoblastic leukemia. Lancet II (7453) 27—28 (1966).

132. —, Y. NAJEAN, M. WEIL, J. TANZER, M. BOIRON et J. BERNARD: Traitement des leucémies aiguës lymphoblastiques par la rubidomycine. Colloque international sur la rubidomycine et la daunomycine. Paris, Hôpital Saint-Louis, 11 mars 1967. Pathol. Biol. **15**, 913—917 (1967).

133. —, M. WEIL et M. BOIRON: Effets de la méthode de réinduction au cours du traitement des leucémies aiguës lymphoblastiques. Nouvelle Rev. Franç. Hématol. **7**, 677—682 (1967) et Actualités hématologiques. II. Masson éd., Paris (1968). A paraître.

134. — — —, and J. BERNARD: On the value of "reinduction" treatments for the control of acute lymphoblastic leukemia (ALL). XIth Congress of the Int. Soc. of Haematol., Sydney (August 1966), 229 (CE 3).

135. — — — — Traitement des leucémies aiguës par la rubidomycine (13057 R.P.). Vth International Congress of Chemotherapy, Wien (Austria) **3**, 289—292 (1967). Verlag der Wiener Medizinischen Akademie (1967).

136. — —, Y. NAJEAN, J. TANZER, P. LORTHOLARY, M. BOIRON et J. BERNARD: Un nouvel agent actif dans le traitement des leucémies aiguës: la rubidomycine (13057 R.P.). Arzneimittel-Forsch. **17**, 955—959 (1967).

137. JAMES, K. W., and H. E. M. KAY: Some aspects of current therapy in acute leukemia. Lancet I (7483) 206—209 (1967).

138. JERNE, N. K., and A. A. NORDIN: Plaque formation in agar by single antibody-producing cells. Science **140**, 405 (1963).

139. JULOU, L., R. DUCROT, J. FOURNEL, P. GANTER, R. MARAL, P. POPULAIRE, F. KOENIG, J. MYON, S. PASCAL et J. PASQUET: Un nouvel antibiotique doué d'activité antitumorale: la rubidomycine (13057 R.P.). III. Etude toxicologique et pharmacologique. Arznei-mittel-Forsch. **17**, 948—954 (1967).

140. KARNOFSKY, D. A.: Meaningful clinical classification of therapeutic responses to anti-cancer drugs. Clin. Pharmacol. Therap. **2**, 709—712 (1961).

141. KARON, M.: Preliminary report on vincristine (Oncovin) from acute leukemia group B. Proc. Amer. Ass. Cancer Res. **4**, 35, abstr. 107 (1963).

142. —, E. FREIREICH, and P. CARBONE: Effective combination therapy of adult acute leukemia. Proc. Amer. Ass. Cancer Res. **6**, 34, abstr. 133 (1965).

143. KAY, E. R. M., N. S. SIMMONS, and A. L. DOUNCE: An improved preparation of sodium desoxyribonucleate. J. Amer. chem. Soc. **74**, 1724—1726 (1952).

144. KERSTEN, W., u. H. KERSTEN: Die Bindung von Daunomycin, Cinerubin und Chromo-mycin A_3 an Nucleinsäuren. Biochem. Z. **341**, 174—183 (1964—1965).

145. —, and W. SZYBALSKI: Physicochemical properties of complexes between deoxyribo-nucleic acid and antibiotics which affect ribonucleic acid synthesis (actinomycin, daunomycin, cinerubin, nogalamycin, chromomycin, mithramycin and olivomycin). Biochemistry **5**, 236—244 (1966).

146. KING, E. J.: The colorimetric determination of phosphorus. Biochem. J. **26**, 292—297 (1932).

147. KLEINKNECHT, D., CL. JACQUILLAT, M. WEIL, Y. NAJEAN, M. BOIRON et J. BERNARD: Les accidents neurologiques centraux de la vincristine. Nouv. Rev. franç. Hémat. **7**, 132—136 (1967).

148. KOSCHEL, K., G. HARTMANN, W. KERSTEN u. H. KERSTEN: Die Wirkung des Chromo-mycins und einiger Anthracyclinantibiotica auf die DNA-abhängige Nuclensäure-Synthese. Biochem. Z. **344**, 76—86 (1966).

149. KOURILSKY, F. M., J. M. DUPUY, D. FRADELIZI, CL. JACQUILLAT et J. BERNARD: Abolition des réactions d'hypersensibilité retardée et gravité des aplasies induites par la rubido-mycine. Colloque international sur la rubidomycine et la daunomycine. Paris, Hôpital Saint-Louis, 11 mars 1967. Pathol. Biol. **15**, 959 (1967).

150. LEMERLE, J., et O. SCHWEISGUTH: Essai de la rubidomycine sur onze cas de tumeurs solides de l'enfant. Colloque international sur la rubidomycine et la daunomycine. Paris, Hôpital Saint-Louis, 11 mars 1967. Pathol. Biol. **15**, 971—972 (1967).

151. LEVACHER, A., J. TANZER, F. M. KOURILSKY et CL. JACQUILLAT: Les transfusions de globules blancs. Les Actualités hématologiques. II. Masson éd., Paris (1968). A paraître.

152. LEVY, J. P., et P. LORTHOLARY: Cytologie pratique du sang et des organes hémato-poïétiques. Doin éd., Paris, 267 p. (1966).

153. LORTHOLARY, P., J. BONHOMME, CL. JACQUILLAT, F. MIELOT, J. BRIERE, M. BOIRON et J. BERNARD: Evolution cytologique des leucémies aiguës lymphoblastiques traitées par la rubidomycine (note préliminaire). Nouv. Rev. franç. Hémat. **7**, 130—132 (1967).

154. — —, F. MIELOT, CL. JAQUILLAT et M. BOIRON: L'évolution cytologique des leucémies aiguës traitées par la rubidomycine. Colloque international sur la rubidomycne et la daunomycine. Paris, Hôpital Saint-Louis, 11 mars 1967. Pathol. Biol. **15**, 945 (1967).

155. LOUBATIERES, A.: Méthodes d'étude et pharmacodynamie des substances cardiotoniques. J. Physiol. (Paris) **43**, 517—580 (1951).

156. LOWRY, O. H., N. R. ROBERTS, and J. I. KAPPHAHN: The fluorometric measurement of pyridine nucleotides. J. Biol. Chem. **224**, 1047—1064 (1957).

157. —, N. J. ROSEBROUGH, A. L. FARR, and R. J. RANDALL: Protein measurement with the Folin phenol reagent. J. Biol. Chem. **193**, 265—275 (1951).

158. LYASHENKO, V. A., and L. P. KOLESNIKOVA: The effect of antitumor antibiotics on pro-duction of hemolysins (russ.). Antibiotiki **12**, 312—315 (1967).

159. MACREZ, CL., H. MARNEFFE-LEBREQUIER, J. RIPAULT, J. P. CLAUVEL, CL. JACQUILLAT et M. WEIL: Accidents cardiaques observés au cours des traitements des leucémies par la rubidomycine. Colloque international sur la rubidomycine et la daunomycire. Paris, Hôpital Saint-Louis, 11 mars 1967. Pathol. Biol. **15**, 949—953 (1967).

160. MAEVSKII, M. M.: Effect of the antibiotic rubomycin on experimental tumor (russ.). Vopr. Protivorak. Bor'by, Vilnyus, Sb, 188—189 (1964).

161. —, E. A. ROMANENKO, A. S. BONDAREVA, A. P. URAZOVA, I. A. AVDEEVA, V. G. MAZAEVA, E. A. TIMOFEEVSKAYA, L. A. SEDAKOVA, Yu. N. MOLKOV, V. A. TALYZINA, and V. V. SILANTIEV: The sensitivity of animal transplantable tumors to rubomycin (russ.). Antibiotiki **12**, 315—319 (1967).

162. MAGLIULO, E., G. STASSANO, and G. P. FIORI: Antimitotic activity of daunomycin. Haematologica (Pavia) **49**, 1091—1102 (1964).

163. MANAPHOVA, N. A.: Conditions for rubomycin production by Actinomyces coeruleo-rubidus (russ.). Antibiotiki **11**, 872—877 (1966).

164. MARAL, R., G. BOURAT, R. DUCROT, J. FOURNEL, P. GANTER, L. JULOU, F. KOENIG, J. MYON, S. PASCAL, J. PASQUET, P. POPULAIRE, Y. DE RATULD et G. H. WERNER: Etude toxicologique et activité antitumorale expérimentale de la rubidomycine (13057 R.P.). Colloque international sur la rubidomycine et la daunomycine. Paris, Hôpital Saint-Louis, 11 mars 1967. Pathol. Biol. **15**, 903—908 (1967).

165. — —, J. FOURNEL, P. GANTER, Y. DE RATULD et G. H. WERNER: Un nouvel antibiotique doué d'activité antitumorale: la rubidomycine (13057 R.P.). II. Activité antitumorale expérimentale. Arzneimittel-Forsch. **17**, 939—948 (1967).

166. MARSH, J. P., JR., C. W. MOSHER, E. M. ACTON, and L. GOODMAN: The synthesis of daunosamine. Chem. Comm. 973—975 (1967).

167. MASSIMO, L.: Premiers résultats obtenus avec la daunomycine dans la thérapie de la leucémie aiguë lymphoblastique et des tumeurs malignes de l'enfant. Communication au Colloque international sur la rubidomycine et la daunomycine. Paris, Hôpital Saint-Louis, 11 mars 1967.

168. MASSIMO, I.: La terapia della leucemia del bambino. Minerva Pediat. **19**, 446 (1967).

169. —, A. FOSSATI-GUGLIELMONI e E. FORTUNA: Primi resultati sull' efficacia terapeutica nella leucemia e nei tumori maligni del bambino di un nuovo antiblastico antibiotico, la „daunomicina". Tumori **10**, 3—19 (1967).

170. —, P. G. MORI, F. COTTAFAVA e E. FORTUNA: Ruolo della chemioterapia antiblastica sul decorso della leucemia del bambino. Valutazione della casistica della Clinica Pediatrica „G. Gaslini" degli anni 1955—1966. Accademia Medica **82** (1967). In stampa.

171. MATHÉ, G., J. L. AMIEL, M. HAYAT, L. SCHWARZENBERG, M. SCHNEIDER, A. CATTAN, J. R. SCHLUMBERGER et B. CEOARA: Essai de traitement des leucémies aiguës et de la maladie de Hodgkin par la rubidomycine (ou daunomycine) seule ou en association. Semaine Hôp. Paris **43**, 2108—2114 (1967).

172. —, M. HAYAT, L. SCHWARZENBERG, M. SCHNEIDER, A. CATTAN, J. R. SCHLUMBERGER et J. L. AMIEL: Essai de traitement des leucémies aiguës par la rubidomycine (ou dauno-mycine) seule ou en association. Colloque international sur la rubidomycine et la daunomycine. Paris, Hôpital Saint-Louis, 11 mars 1967. Pathol. Biol. **15**, 933—937 (1967).

173. —, L. SCHWARZENBERG, M. SCHNEIDER, J. L. AMIEL, A. CATTAN et J. F. SCHLUMBERGER: Etude sur l'évolution et le traitement de la leucémie aiguë lymphoblastique. Bull. Mém. Soc. Méd. Hôp., Paris **117**, 711—734 (1966).

174. — — —, J. R. SCHLUMBERGER, M. HAYAT, J. L. AMIEL, A. CATTAN, and C. JASMIN: Acute lymphoblastic leukaemia treated with a combination of prednisone, vincristine, and rubidomycin. Value of pathogen-free rooms. Lancet II (7512), 380—382 (1967).

175. OLLIS, W. D., and I. O. SUTHERLAND: A new family of antibiotics. In OLLIS, W. D.: Chemistry of natural phenolic compounds, p. 212 Oxford: Pergamon Press (1961).

176. OYAMA, V. I., and H. EAGLE: Measurement of cell growth in tissue culture with a phenol reagent (Folin-Ciocalteau). Proc. Soc. exp. Biol. Med. **91**, 305—307 (1956).

177. PARISI, B., and A. SOLLER: Studies on the antiphage activity of daunomycin. Giorn. Microbiol. **12**, 183—194 (1964).

178. PEVZNER, N. S., and S. P. SHAPOVALOVA: Study of antimicrobial properties and effect of rubomycin on intestine microflora of experimental animals (russ.). Antibiotiki **12**, 523—526 (1967).

179. PIERCE, M. I.: The acute leukemia of childhood. Pediat. Clin. N. Amer. **4**, 497—530 (1957).

180. PREOBRASHENSKAYA, T. P., N. A. MANAPHOVA, and G. F. GAUSE: Systematic position, variation and antibiotic properties of rubomycin producing organism (russ.). Anti-biotiki **11**, 867—872 (1966).

181. PRICE, K. E., R. E. BUCK, and J. LEIN: Incidence of antineoplastic activity among anti-biotics found to be inducers of lysogenic bacteria. Antimicrobial Agents and Chemo-therapy 505—517 (1964).

182. PÜTTER, J.: Über die biochemischen Wirkungen von Cytostatica auf Tumorzellen. In Therapie maligner Tumoren, Bd. I. Hrsg. von F. MEYTHALER. Stuttgart: F. Enke (1966).

183. RATULD, Y. DE, et G. H. WERNER: Mise en évidence dans des cultures de cellules em-bryonnaires de caille d'un virus latent, cytopathogène, tumorigène pour la caille et le poussin. Relations entre ce virus et celui du sarcome de Rous. Ann. Inst. Pasteur **113**, 749—755 (1967).

184. RHONE-POULENC, S. A.: French patent specification N⁰ 898076 filed 18/5/62. German patent N⁰ 1215863 (filed 14/5/63, published 29/8/66, French priority date 18/5/62).

185. RICHARDSON, A. C.: The synthesis of D-daunosamine N-benzoate. Chem. Comm., 627—628 (1965).

186. RIPAULT, J., M. WEIL et CL. JACQUILLAT: Etude nécropsique de quatre malades traités par la rubidomycine. Colloque international sur la rubidomycine et la daunomycine. Paris, Hôpital Saint-Louis, 11 mars 1967. Pathol. Biol. **15**, 955—957 (1967).

187. ROSSOLIMO, O. K., and G. N. LEPESHKINA: Experimental study on antitumor effect of rubomycin in combination with some cytostatic drugs (russ.). Antibiotiki **12**, 206—211 (1967).

188. Rusconi, A.: Different binding sites in DNA for actinomycin and daunomycin. Biochim. Biophys. Acta 123, 627—630 (1966).

189. —, and E. Calendi: Action of daunomycin on nucleic acid synthesis in hepatoma ascites cells incubated in vitro. Tumori 50, 261—266 (1964).

190. — — Action of daunomycin on nucleic acid metabolism in HeLa cells. Biochim. Biophys. Acta 119, 413—415 (1966).

191. Sanfilippo, A., e R. Mazzoleni: Attività antifagica dell' antibiotico daunomicina. Giorn. Microbiol. 12, 83—92 (1964).

192. Scarpinato, B., L. Valentini e E. Calendi: Studi sull' interazione tra daunomicina e acidi nucleici. IV Simposio Nazionale della Società Italiana di Cancerologia, Milano, 25—26/10/1963.

193. Schmidt, G., and S. J. Thannhauser: A method for the determination of desoxyribonucleic acid, ribonucleic acid and phosphoproteins in animal tissues. J. Biol. Chem. 161, 83—89 (1945).

194. Schreck, R.: A method for counting the viable cells in normal and in malignant cell suspensions. Amer. J. Cancer 28, 389—392 (1936).

195. Schwarzenberg, L., G. Mathe, J. L. Amiel, A. Cattan, M. Schneider et J. R. Schlumberger: Le traitement symptomatique de l'agranulocytose par les transfusions de globules blancs. Presse méd. 74, 1057—1060 (1966).

196. Selawry, O. S., and E. Frei III: Prolongation of remission in acute lymphocytic leukemia by alteration in dose schedule and route of administration of methotrexate. Clin. Res. 12, 231 (1964).

197. Serpick, A. A., and E. S. Henderson: Observations on toxicity and clinical trials with daunomycin. Colloque international sur la rubidomycine et la daunomycine. Paris, Hôpital Saint-Louis, 11 mars 1967. Pathol. Biol. 15, 909—912 (1967).

198. Shorine, V. A., O. C. Rossolimo, L. E. Goldberg, M. S. Stanislavskaya et N. A. Blumberg: Activité antitumorale et antivirale de la rubomycine. International IX Congress for Microbiology, Moscow, 24—30/7/1966. Abstracts of Papers, p. 167. Organizing Committee of the Congress, Moscow (1966).

199. — —, M. S. Stanislavskaya, N. A. Blumberg, and G. N. Lepeshkina: An experimental study of rubomycin antitumor activity (russ.). Antibiotiki 11, 14—20 (1966).

200. —, and S. P. Shapovalova: Selection of new natural inhibitors of immunogenesis (russ.). Antibiotiki 11, 963—967 (1966).

201. Silvestrini, R., A. di Marco, S. di Marco e T. Dasdia: Azione della daunomicina sul metabolismo degli acidi nucleici di cellule normali e neoplastiche coltivate in vitro. Tumori 49, 399—412 (1963).

202. — e M. Gaetani: Azione della daunomicina sul metabolismo nucleico del tumore ascite di Ehrlich. Tumori 49, 389—398 (1963).

203. Simard, R.: Specific nuclear and nucleolar ultrastructural lesions induced by proflavin and similarly acting antimetabolites in tissue culture. Cancer Res. 26 (Pt 1), 2316—2328 (1966).

204. Skipper, H. E.: Perspectives in cancer chemotherapy: therapeutic design. Cancer Res. 24, 1295—1302 (1964).

205. —, and L. H. Schmidt: A manual on quantitative drug evaluation in experimental tumor systems. Part I—Background, description of criteria, and presentation of quantitative therapeutic data on various classes of drugs obtained in diverse experimental tumor systems. Cancer Chemotherapy Rept. 17, 1—143 (1962).

206. Stanislavskaya, M. S., and T. P. Vertogradova: The effect of rubomycin on peripheral blood of rabbits and dogs (russ.). Antibiotiki 11, 691—695 (1966).

207. Sternberg, S. S., and F. S. Philips: Biphasic intoxication and nephrotic syndrome in rats given daunomycin. Proc. Amer. Ass. Cancer Res. 8, 64, abstr. 252 (1967)

208. Strelitz, F., H. Flon, U. Weiss, and I. N. Asheshov: Aklavin, an antibiotic substance with antiphage activity. J. Bacteriol. 72, 90—94 (1956).

209. Sultan, Y.: La transfusion de plaquettes chez les sujets en aplasie thérapeutique. Les Actualités hématologiques. II. Paris: Masson 1968. A paraître.

210. Tan, C.: Childhood leukemia. An antibiotic curbs course. Med. Tribune 7, 1 (1966).

211. TAN, C., R. B. GOLBEY, C. L. YAP, N. WOLLNER, C. A. HACKETHAL, M. L. MURPHY, H. W. DARGEON, and J. H. BURCHENAL: Clinical experiences with actinomycins D, KS 2 and F 1 (KS 4). Ann. N. Y. Acad. Sci. **89**, 426—444 (1960).

212. —, and H. TASAKA: Daunomycin remissions in acute leukemia. Proc. Amer. Ass. Cancer Res. **7**, 70, abstr. 278 (1966).

213. — —, and A. DI MARCO: Clinical studies of daunomycin. Proc. Amer. Ass. Cancer Res. **6**, 64, abstr. 253 (1965).

214. — —, YU. KOU-PING, M. L. MURPHY, and D. A. KARNOFSKY: Daunomycin, an antitumor antibiotic, in the treatment of neoplastic disease. Clinical evaluation with special reference to childhood leukemia. Cancer **20**, 333—353 (1967).

215. TANZER, J., et M. BOIRON: Essais de nouveaux traitements dans la leucémie myéloïde chronique. Entretiens de Bichat, 1967. Thérapeutique. Paris, Expansion Scientifique Française, ed., 115—117 (1967). Semaine Therap., Sem. Hôp. **44**, 402—404 (1968).

216. — —, CL. JACQUILLAT, M. WEIL, D. LEVY et J. BERNARD: Effets de la rubidomycine dans la leucémie myéloïde chronique. Colloque international sur la rubidomycine et la daunomycine. Paris, Hôpital Saint-Louis, 11 mars 1967. Pathol. Biol. **15**, 943—944 (1967).

217. —, CL. JACQUILLAT, M. WEIL, D. LEVY, Y. NAJEAN, M. BOIRON et J. BERNARD: Effets de l'hydroxyurée dans la leucémie myéloïde chronique. Etude préliminaire. Presse méd. **74**, 2929—2930 (1966).

218. THEOLOGIDES, A., J. W. YARBRO, and B. J. KENNEDY: Daunomycin inhibition of DNA and RNA synthesis in normal and malignant tissues. Proc. Amer. Ass. Cancer Res. **8**, 67, abstr. 264 (1967).

219 a. THOMAS, M. A.: L'antibiothérapie dans le traitement des aplasies dans les leucémies. Les Actualités hématologiques. II. Paris: Masson 1968. A paraître.

219 b. TONG, G. L., P. LIM, and L. GOODMAN: Identity of rubidomycin and daunomycin. J. Pharm. Sci. **56**, 1691—1692 (1967).

220. TRUHAUT, R., et G. DEYSSON: Etude, sur le test Allium, des propriétés antimitotiques de la daunomycine. C. R. Soc. Biol. **160**, 283—285 (1966).

221. TURSUNKHODZHAEV, N. B.: The mechanism of action of actinomycin C antibiotic on the tumor cell (russ.). Antibiotiki **8**, 111—114 (1963).

222. UMBREIT, W. W., R. H. BURRIS, and J. F. STAUFFER: Manometric techniques. Minneapolis: Burgess Publishing Co. 1959.

223. UPJOHN: U.S. Patent N° 3183157 (filed 1/2/63, published 11/5/65).

224. VENDITTI, J. M., B. J. ABBOTT, A. DI MARCO, and A. GOLDIN: Effectiveness of dauno-mycin (NSC-82151) against experimental tumours. Cancer Chemotherapy Rept. **50**, 659—665 (1966).

225. —, I. KLINE, and A. GOLDIN: Evaluation of antileukemic agents employing advanced leukemia L 1210 in mice. VIII. Cancer Res., Suppl. **24** (Part 2), 827—1091 (1964).

226. WAKSMAN, S. A.: The actinomycins and their importance in the treatment of tumors in animals and man. Ann. N. Y. Acad. Sci. **89**, 285—485 (1960).

227. —, and H. B. WOODRUFF: Bacteriostatic and bactericidal subtances produced by a soil actinomyces. Proc. Soc. exp. Biol. Med. **45**, 609—614 (1940).

228. WARD, D. C., E. REICH, and I. H. GOLDBERG: Base specificity in the interaction of polynucleotides with antibiotic drugs. Science **149**, 1259—1263 (1965).

229. WEIL, M., CL. JACQUILLAT et M. BOIRON: Traitement des lymphosarcomes par la triple association vincristine—rubidomycine—prednisone. Colloque international sur la rubi-domycine et la daunomycine. Paris, Hôpital Saint-Louis, 11 mars 1967. Pathol. Biol. **15**, 967—968 (1967).

230. — — — et J. BERNARD: Etude d'une quadruple association chimiothérapique dans le traitement des leucémies aiguës granulocytaires. Europ. J. Cancer A paraître.

231. WEITZEL, G., u. E. BUDDECKE: Vergleich der Einwirkung von Zink, Persauerstoff und Alkylierungsmitteln auf Ascites-Zellen. Hoppe-Seylers Z. Physiol. Chem. **323**, 14—30 (1961).

232. WHEELER, G. P.: Studies related to the mechanisms of action of cytotoxic alkylating agents: a review. Cancer Res. **22**, 651—688 (1962).

233. ZELEZNICK, L. D., and C. M. SWEENEY: Inhibition of deoxyribonuclease action by nogalamycin and U-12241 by their interaction with DNA. Arch. Biochem. Biophys. **120**, 292—295 (1967).

234. ZUELZER, W.: Implications of long-term survival in acute stem cell leukemia of childhood treated with composite cyclic therapy. Blood **24**, 477—494 (1964).

Addendum

235. ARCAMONE, F., G. CASSINELLI, G. FRANCESCHI, and P. OREZZI: The total absolute configuration of daunomycin. Tetrahedron Letters (30) 3353—3356 (1968).

236. —, G. FRANCESCHI, P. OREZZI, and S. PENCO: The structure of daunomycin. Tetrahedron Letters (30) 3349—3352 (1968).

237. ASTRAKHAN, V. I.: Clinical use of antitumor antibiotics (russ.). Antibiotiki **12**, 157—172 (1967).

238. BERNARD, J.: Acute leukemia treatment. Cancer Res. **22**, Part I, 2565—2569 (1967).

239. BERNSTEIN, J. G., J. C. CRADOCK, W. L. THOMPSON, and N. R. BACHUR: Cardiovascular pharmacology of daunomycin. Clin. Res. **16**, 357 (1968).

240. BONDAREVA, A. S., and M. M. MAEVSKII: The mechanism of action of some new antitumor antibiotics (russ.). Antibiotiki **13**, 208—212 (1968).

241. BRAHIC, M.: Effets de deux inhibiteurs de la synthèse des RNA, l'actinomycine D et la rubidomycine, sur la replication d'un virus à RNA: le virus Sindbis. Thése de Médecine, Marseille, 29 juin 1968 (polycopie).

242. CHAPTAL, J., R. JEAN, P. IZARN, H. BONNET et M. NAVARRO: Indications de la rubidomycine dans les leucoses aiguës. Modifications cytologiques au cours de traitement. Ann. Pédiat Sem. Hôp., **44**, 1842 (1968).

243. DI MARCO, A., C. BORETTI, A. RUSCONI, and R. SILVESTRINI: Metabolic degradation and antineoplastic activity of daunomycin. Abstracts of Papers, Ninth International Cancer Congress, p. 376 (1966). International Union against Cancer, Tokyo, 23—29 October 1966, in Cancer Chemother. Abstr. **7**, abstr. 3155 (1966).

244. —, E. CALENDI, G. DI FRONZO, and A. SCIACCHITANO: Synergistic effect of daunomycin and immunization on the growth of transplantable tumors (ital.). Tumori **53**, 453—459 (1967).

245. DOROZHINSKII, V. B.: Comparative in vitro studies on the interaction of rubomycin B₂ and C with DNA (russ.). Antibiotiki **13**, 198—201 (1968).

246. DUPUY, J. M., F. M. KOURILSKY, D. FRADELIZI, J. DAUSSET et J. BERNARD: Abolition des réactions d'hypersensibilité retardée et gravité des aplasies induites par la rubidomycine. Nouv. Rev. franç. Hémat. **7**, 907 (1967).

247. FEO, C.: Action de la rubidomycine sur l'hématopoïèse de la souris. C. R. Soc. Biol. **161**, 1864—1866 (1967).

248. GAUSE, G. F.: Study of the mechanism of action of antitumor antibiotic rubomycin. Abstracts of Papers, Ninth International Cancer Congress, p. 353 (1966). International Union against Cancer, Tokyo, 23—29 October 1966, in Cancer Chemother. Abstr. **7**, abstr. 2276 (1966).

249. GOUDEMAND, M., D. SAUTIERE-HABAY et Y. DELMAS-MARSALET: La rubidomycine dans le traitement de la leucémie aiguë (32 cas). Lille Méd. **12**, 1163—1170 (1967).

250. GROUCHY, J. DE, et C. DE NAVA: Effets cytogénétiques de la rubidomycine (ou daunomycine). Ann. Génét. **11**, 39—44 (1968).

251. GUYON, J. M., D. FIERE, L. REVOL et P. CROIZAT: Six mois d'expérience de la rubidomycine dans le traitement des leucoses aiguës. Nouv. Rev. franç. Hémat. **8**, 490 (1968).

252. HERMAN, E. H.: Study of the cardiac toxicity of the antibiotic daunomycin (NSC 82151) in rodents. Fed. Proc. **27**, 707, abstr. 2757 (1968).

253. —, P. SCHEIN, and R. M. FARMER: Pharmacological modification of the daunomycin induced arrhythmia in the hamster. Pharmacologist **10**, 220, abstr. 376 (1968).

254. IWAMOTO, R. H., P. LIM, and N. S. BHACCA: The structure of daunomycin. Tetrahedron Letters (36) 3891—3894 (1968).

255. JACQUILLAT, CL., M. WEIL, M. BOIRON et J. BERNARD: Traitement des leucémies aiguës par la rubidomycine (13 057 R.P.). Un. Méd. Can. **97**, 8—12 (1968).

256. Kessel, D., V. Botterill, and I. Wodinsky: Uptake and retention of daunomycin by mouse leukemic cells as factors in drug response. Cancer Res. **28**, 938—941 (1968).
257. Khuskivadze, B. K.: Effect of antitumor antibiotics on synthesis of RNA and proteins in rabbit lymphocytes (russ.). Antibiotiki **13**, 340—344 (1968).
258. Mahour, G. H., E. H. Soule, S. D. Mills, and H. B. Lynn: Rhabdomyosarcoma in infants and children: a clinico-pathologic study of 75 cases. J. Pediat. Surg. **2**, 402—409 (1967).
259. Malpas, J. S., and R. B. Scott: Rubidomycin in acute leukaemia in adults. Brit. med. J. **3**, 227—229 (1968).
260. Marsh, J. P. Jr., and L. Goodman: The synthesis of some aminotrideoxy sugars related to daunosamine. Amer. Chem. Soc. 155th Meet., San Francisco, March 31—April 5 (1968). Abstracts of Papers, Division of Carbohydrate chemistry, C 15.
261. —, R. H. Iwamoto, and L. Goodman: Synthesis and characterisation of compounds related to daunomycin. Chem. Comm. 589—590 (1968).
262. Massimo, I., A. Fossati-Guglielmoni, and E. Fortuna: Remission in acute leukaemia. Brit. med. J. **4**, 54 (1967).
263. Matthews, R. N.: Daunomycin/Rubidomycin: a new anti-leukaemic drug. Aust. Ann. Med. **17**, 176 (1968).
264. Mazaeva, V. G.: On the problem of accelerated growth of transplantable tumors in animals previously treated with antitumor antibiotics (russ.). Antibiotiki **13**, 212—215 (1968).
265. Ribas-Mundo, M.: New method and drugs in the treatment of leukemia. (span.). Med. Clin. (Barcel.) **48**, 245—247 (1967).
266. Rusconi, A., G. di Fronzo, and A. di Marco: Distribution of tritiated daunomycin (NSC 82 151) in normal rats. Cancer Chemotherapy Rept. **52**, 331—335 (1968).
267. Scott, R. B.: Acute leukaemia. The current therapeutic position. Proc. roy. Soc. Med. **61**, 471—476 (1968).
268. Segni, G., R. Mastrangelo, and G. Tortorolo: Daunomycin in Letterer-Siwe's disease. Lancet **1968** II, 461.
269. Soulie, J.: Un nuevo antibiotico dotado de actividad antitumoral: la rubidomicina. Rev. Clin. Espan. **108**, 499—502 (1968).
270. Staeuber, P. G., u. H. Gerhartz: Die Chemotherapie mit Daunomycin bei Lymphogranulom und Retikulosarkomkranken. Verh. dtsch. Ges. inn. Med. **73**, 334—339 (1967).
271. Theologides, A., J. W. Yarbro, and B. J. Kennedy: Daunomycin inhibition of DNA and RNA synthesis. Cancer **21**, 16—21 (1968).
272. Vig, B. K., S. B. Kontras, E. F. Paddock, and L. D. Samuels: Daunomycin induced chromosomal aberration and the influence of arginine in modifying the effect of the drug. Mutat. Res. **5**, 279—287 (1968).
273. — —, and L. D. Samuels: Chromosome aberrations induced by daunomycin in human leucocyte cultures, with the apparent synergistic effect of arginine. Experientia **24**, 271—273 (1968).
274. Young, C. W.: Cancer chemotherapy: new agents and new applications of established agents. Acad. Med. New Jersey Bull. **13**, 46—59 (1967).

Subject Index

Herstellung: Konrad Triltsch, Graphischer Betrieb, Würzburg

Monographs already Published

In Preparation

In Preparation

ACKERMANN, N. B., Boston: Use of Radioisotopic Agents in the Diagnosis of Cancer

Asparaginase. Edited by E. GRUNDMANN, Wuppertal-Elberfeld, and OETTGER, New York (Symposium)

BOIRON, M., Paris: The Viruses of the Leukemia-sarcoma Complex

CAVALIERE, R., A. ROSSI-FANELLI, B. MONDOVI, and G. MORICCA, Roma: Selective Heat Sensitivity of Cancer Cells

CHIAPPA, S., Milano: Endolymphatic Radiotherapy in Malignant Lymphomas

Cutane paraneoplastische Syndrome. Edited by J. J. HERZBERG, Bremen (Symposium)

DENOIX, P., Villejuif: Le traitement des cancers du sein

GRUNDMANN, E., Wuppertal-Elberfeld: Morphologie und Cytochemie der Carcinogenese

IRLIN, I. S., Moskva: Mechanisms of Viral Carcinogenesis

LANGLEY, F. A., and A. C. CROMPTON, Manchester: Epithelial Abnormalities of the Cervix Uteri

MATHÉ, G., Villejuif: L'Immunotherapie des Cancers

MEEK, E. S., Bristol: Antitumour and Antiviral Substances of Natural Origin

NEWMAN, M. K., Detroit: Neuropathies and Myopathies Associated with Occult Malignancies

OGAWA, K., Osaka: Ultrastructural Enzyme Cytochemistry of Azo-dye Carcinogenesis

PARKER, J. W., and R. J. LUKES, Los Angeles: Lymphocyte Transformation

PENN, I., Denver: Malignant Lymphomas in Transplant Patients

Recent Advances in the Treatment of Acute Leukemias. Edited by G. MATHÉ (Symposium)

SUGIMURA, T., Tokyo, H. ENDO, Fukuoka, and T. ONO, Tokyo: Chemistry and Biological Action of 4-Nitroquinoline 1-oxide, a Carcinogen

WEIL, R., Lausanne: Biological and Structural Properties of Polyoma Virus and its DNA

WILLIAMS, D. C., Caterham, Surrey: The Basis for Therapy of Hormon Sensitive Tumours

WILLIAMS, D. C., Caterham, Surrey: The Biochemistry of Metastasis